Ligang Xiang

The 5G Era

What is 5G and How Will it Change the World?

palgrave
macmillan

Ligang Xiang
Information Consumption Alliance
Beijing, China

ISBN 978-981-99-4533-7 ISBN 978-981-99-4534-4 (eBook)
https://doi.org/10.1007/978-981-99-4534-4

Jointly published with China Renmin University Press

This Palgrave Macmillan imprint is published by the registered company Springer Nature Singapore Pte Ltd.
The registered company address is: 152 Beach Road, #21-01/04 Gateway East, Singapore 189721, Singapore

Preface

The Door to 5G Has Just Opened

In October 2018, a tragedy occurred in Wanzhou, Chongqing: A bus crashed into a car going in the opposite direction and plunged 70 metres off a bridge, killing everyone on board. Such a vicious incident led to widespread discussion and speculation on the Internet, from blaming the female driver of the car that was hit in the opposite direction, to speculating on the psychology of the driver. At that time, some people also said that we might never know the truth, because everyone on the bus was gone.

However, we soon learned the truth. When the car was recovered, a camera on the car recorded the scene, and the footage shows that the tragedy was caused by someone arguing with the driver and the driver hitting the steering wheel. The truth was revealed because of the power of technology, namely the car recorder on the bus, which recorded everything.

I was thinking, if with 5G, the bus had to run along a digital track, and when the control centre found it was off the track it should be on, it could be the first to send a control command over a low latency network to brake the bus, and also to brake other buses also running on this digital track, then a dozen lives might not have been lost. I am sure this will happen.

Every time there is an upgrade in mobile technology, there is a voice that says, "Do consumers want it? When 3G arrived, there were many voices saying that consumers could use their mobile phones to make phone calls and send text messages, not to access the Internet. When 4G arrived, there were many people who thought that 3G bandwidth was enough and the investment had not yet been recovered, so why should we engage in 4G?

5G is not just a technology, it is a world-changing force through technology.

When the Internet arrives and we discuss how to bridge the digital divide, if only to discuss that it will neither allow the Internet to reach remote areas nor allow the poor to afford computers and learn how to use and operate them. Many charity projects are only paper and superficial.

With the arrival of 4G, the "Village to Village" project has enabled most areas in China to be covered by 4G networks, with cheap tariffs and cheap smartphones, allowing ordinary people in remote areas to enter the Internet era overnight. Social, e-commerce and mobile payments are applications that were once considered to be used by the high end of the population and have entered the lives of ordinary people. Everyone in the village is in the same group, sharing everything from celebrations to job vacancies, and even discussing how to make beef dumplings, This is where the digital divide is truly bridged, and this is done through the underlying power of the network.

This is a book about 5G, but the focus is not on the technology of 5G, as there are already many books that explain 5G technology, and I am not a technical expert myself. This book is an exploration of how the industry will evolve and change under a new network system, and how 5G will impact society and the economy.

Not long ago, after an event, I had noodles with a well-known economist from one of China's top universities, and I was talking about the six information revolutions that have changed the course of humanity, and the impact of the seventh information revolution based on 5G on society as a whole. I was surprised to learn that many people today are out of touch with many aspects of society, economics and technology, and have been for a long time, many Chinese economists are copying economic theories from a hundred years ago or even further back and applying them to China today. An important reason for their lack of innovative exploration of economic theory is their detachment from social practice and their lack of research into the psychological changes in society and the changes in economic relations brought about by technological change and the increase in the productive capacity of society. The world has changed in terms of technology and in terms of economic capacity, and economic theory should also change with the times.

Infrastructure development, industrial cluster capacity and major technological change all have a significant impact on productivity and, ultimately, on the economy, society and culture. 5G, the latest development in information technology capabilities in human history, will have a profound impact on society as a whole. In addition to providing basic human-to-human communication functions, it will extend communication capabilities to humans and machines and become the foundation of the smart Internet.

Therefore, in addition to technology, we need to explore industries, businesses, applications, and even economies and cultures, which, I believe, will be enhanced to influence the social fabric and even the future of humanity.

The world is integrated and complex. Today, we often talk about the Internet of Things, artificial intelligence and big data, but without a strong, efficient communications capability that can provide support anywhere, anytime, these high profile technologies will not progress for many years, or are unlikely to enter the lives of ordinary people. Once the communication capabilities are solved and integrated with these capabilities, a whole new network system will be formed, which is the Smart Internet. The Smart Internet is not the traditional Internet, but with the Internet as its foundation, it will bring about greater and more revolutionary changes to mankind.

The term “Smart Internet” is a concept we put forward together when we talked about the industry with Yang Xu, President of Intel Greater China, almost five years ago. At that time, we didn’t feel able to distinguish between traditional Internet, mobile Internet and smart Internet. Today this division is becoming clearer and clearer, as the traditional Internet is based on the PC as the core terminal, mainly for information transmission, with services as a secondary capability; mobile Internet is a smartphone as the core terminal, mainly for life services, social, mobile payment, location are its core capabilities; The Smart Internet will turn the various devices in our lives into terminals, it will not only provide life services, but will also facilitate public administration and even participate in manufacturing, and we need to clarify step by step what its core competencies are.

It is impossible to cover everything about 5G in this book, but my idea is to take you beyond the technology itself and, based on an understanding of the basic technology and features, to analyse the impact and implications of 5G on all areas of the ecosystem, including the future economic, social and even ideological and cultural implications. Today, the door to 5G has just been opened, and there are many directions that are certainly not clear enough, but I hope this analysis will give you some inspiration.

This book has been shaped by the many insights I have received from a large number of experts and professionals during the hundreds of lectures I have given to the Chinese telecommunications industry. The intensive communication with professionals from companies such as Intel, Ericsson, Nokia, Huawei, ZTE and Datang, and the systematic training on 5G at companies like Qualcomm, has given me more knowledge and understanding of 5G. In addition, there were many exchanges with the Ministry of Industry and Information Technology, telecom operators and leaders of relevant research institutes, and much of the industry information in the book comes from communication with them.

I wrote most of the book myself, while parts of it were written by Li Li, Yang Liang, Wang Jierui and Li Ming after several lengthy interviews, and then edited by me. Thanks to several people who participated in the writing of this book. Of course, special thanks go to Ms. Ma Yue and Mr. Yao Maodun of Koala Look, a premium content operator, for their conscientious and responsible efforts to make this book possible. Finally, I would like to thank the editors of the People's University of China Press.

We hope this book will open up a whole new world for you!

Contents

1

5G as the Basis for the Seventh Information Revolution

The Development of Information Technology Transforms Humanity

The ability to access information is a miracle of human ingenuity and a fundamental force that has transformed humanity.

When an ape is still an ape, it is not essentially different from a lamb. But apes eventually evolved into humans, in large part because they have a different ability to access information than other animals.

Information is also exchanged between lower animals, but in a very homogeneous way: they rely on simple syllables and body language, and therefore convey very little information. The inability of animals to transmit complex information to each other, especially to express their emotions, is the greatest difference between them and humans. The apes, however, have taken a very different path: from the first crawling to upright walking; Tools were created to form a more complex system of sound transmission in the form of language, which was created to enrich information. With the gradual improvement of the language system, the development of information technology in human society began a revolutionary phase.

L. Xiang, *The 5G Era*, https://doi.org/10.1007/978-981-99-4534-4_1

Language Makes the Human Race

Language makes people, and therefore human society.

The Bible records in the Old Testament, Genesis, that God made a pact with mankind that he would not send any more floods to harm the earth by means of a rainbow, and that mankind has communicated in Adam's language ever since. As Noah's descendants grew, the people began to move eastwards, eventually settling in the land of Shinar, near ancient Babylon.

As the saying goes, "Be prepared for danger in times of peace," and this is when people wonder: is it true that there will be no more flooding? Although God has made a promise, there is no need to pin our hopes on it. After discussion, the humans decided to find another habitat in case disaster struck again and they had nowhere to go, so they worked together to build the famous Tower of Babel (also known as the Tower of Heaven), whose top would lead to the heavens. The construction work went very smoothly as the people spoke the same language and there were no barriers to communication. However, this angered God, and to punish the arrogance of mankind, God disrupted the language of mankind. Without a common language, there was no exchange of information and the people had to disperse and the construction of the Tower of Babel was abandoned.

This is the famous Western allusion to the Tower of Heaven, which shows that the failure of information transmission leads to the failure of human cooperation. The creation of language was the first information revolution in human history. It is well known that human language is not the only language in the world, but that animals also have their own ways of communicating, such as insects and whales. But in the end, it was language, with speech as its key vehicle, that helped mankind conquer the world, and ultimately, it was mankind itself that was special.

A Brief History of Man, by Yuval Heraly, mentions that between 70,000 and 30,000 years ago, as apes evolved, their brains developed new ways of thinking and communicating, leading to a cognitive revolution: A chance genetic mutation has changed the way Homo sapiens' brains are wired internally, allowing them to think in ways never before possible and to communicate in a completely new kind of language." Although animals also have language centres, human beings have evolved in the course of evolution, as their brains have evolved to think and communicate, eventually evolving from lower to higher animals, with more developed intelligence, and thus evolving their own set of social attributes, which is the fundamental reason why human language is different from animal language, and an important factor in human society eventually becoming the dominant force in the world.

Compared to animals, human language has multiple functions and is very flexible and versatile in its use, the most important of which is the sharing of information: If an ape finds a threat from a fierce animal ahead, it will pass this information on to the other apes so that they can take precautions. However, a pig that finds the same situation is unable to clearly communicate the danger to others of its kind because it does not have a more sophisticated system of information transmission that can be shared. This contrast already shows that the survival experience of the lower animals cannot be shared, whereas the ape's knowledge and experience of the world can be shared not only with itself, but also with its own kind. The results of this development are promising: Although one ape's knowledge of the world is limited, if many apes share their knowledge, the information is of great value, and language is the vehicle that helps share information and facilitates the evolution of apes. For the first human apes, the advantages of language are truly magical.

How did language come into being? Until now this has remained an unsolved mystery, the most fatalistic claims being the biblical Tower of Babel and the cognitive revolution mentioned above. For the English language, there were later three different schools of thought by three major linguists, namely the structuralist theory of Ferdinand de Saussure (1857–1913), the transformational generative grammar theory of Avram Noam Chomsky (1928–), and the systemic functional linguistics theory of M.A.K. Halliday (1925–1928), 1925–2018), and M.A.K. Halliday's (1925–2018) systemic functional linguistics.

These linguistic masters have made a significant contribution to the world of linguistics, but there are still unanswered questions which we cannot explore in depth, but which, taken together with the views of these masters and those of our leading linguist, Professor Hu Zhuanglin, suggest that human language is creative, displaced, culturally transmitted, interchangeable and meta-linguistic (i.e. language is used to discuss language itself). Human language is not only capable of transmitting existing information, but also of "discussing fictional matters", i.e. it can be used to imagine the future, to make up stories, etc. These characteristics are sufficient for languages to help people on earth to establish various social activities, work-learning cooperation mechanisms, social science systems, etc.

The earliest sounds did not come from the organisms' vocal organs, but were produced accidentally through the interaction of their body parts with fluids, and when animals heard them, they developed a "practice-like" behaviour to imitate and control the sounds they produced. The vocal apparatus evolved as the animal's control of sound was optimised, and the organ that controls all activity is the animal's brain, of which the human brain is particularly well developed. Compared to lower animals, the human brain can store visual memories as images, forming its own database, and when it needs to output, it can then call up the database in the brain and turn the figurative information into a stream of information in the brain, and finally transmit it to others through language, completing the transmission of linguistic information.

In the early days of phonetics, the language's ability to transmit information did not really play much of a role, as humans communicated only with a few simple words. As primitive humans saw and came into contact with more and more things, existing words became inadequate to meet their expressive needs, and new words began to emerge to refer to a particular object. As new vocabulary emerged and reached the stage of redundancy, sentence structure emerged and the original vocabulary was streamlined and optimised. Although there are now many different means of transmitting information, it is undeniable that the creation of language has provided the basis for the construction of information that allows people to live complex social lives and for the human brain to evolve, optimising its thinking and further refining its behavioural capabilities.

Before the advent of writing, language as the only means of transmitting information relied mainly on hearing, and the vehicle for transmitting language was the human articulatory organ, which emitted sound waves that were transmitted through the air to become sound, and the receiving end was the human auditory organ.

It was with language that man was formally distinguished from the lower animals and became a truly "intelligent being". The creation of language initiated the first information revolution and the construction of human society.

In ancient times, when there was no mobile communication at all, and no electricity at all, the environment for the transmission of information by language was very limited: Both the sender and the receiver of the information must be human, and the transmission must be face-to-face, with strict time and space limitations on the exchange of information. Although the quality of the information conveyed is high, once it becomes remote, problems can arise, even in today's society, as long as the information is passed on by word of mouth, it will almost always end up being out of order.

The tragedies caused by the shortcomings of oral transmission of information were commonplace in slaveholding societies. At the end of the sixteenth century, the English playwright Shakespeare's masterpiece, *Julius Caesar*, was a masterful illustration of the historical tragedy of the ruling class and the aristocracy's control of speech in the context of Roman slavery. When Caesar was on the verge of becoming emperor, his consul and republican representative, Cassius, joined forces with other nobles and persuaded Brutus, a powerful consul close to Caesar, to plan Caesar's assassination. After the success of the operation, Antony, Caesar's confidant, pretended to make amends with the republicans, and Brutus was kind enough to allow Antony to make a speech to the Roman public. In fact, Antony took the opportunity of his speech to incite the Roman public to turn against the republicans and cause unrest. Cassius and Brutus were forced to flee and eventually committed suicide.

In a slave society, where information was not available and the people's minds were suppressed, the agitation of public opinion became a powerful tool for the ruling class to brainwash the masses. When Caesar died by the sword of the republicans, the city was in turmoil and Brutus took to the stage to tell the public how dictatorial and murderous Caesar was and that if he was allowed to claim the throne, Rome would be plunged into a dark reign. Brutus' impassioned tone and righteousness united the panic-stricken Roman people, who chanted "Brutus" and abhorred Caesar's evil deeds, believing that he deserved to die. However, when Brutus leaves, Antony takes to the stage to pay tribute to Caesar, to celebrate the righteous actions of the republicans, and, while appealing to the public, he takes the opportunity to turn the tables on the republicans and unexpectedly begins to tell the story of Caesar's heroic deeds. At a critical moment when the people were at a loss, Antony pulled out a roll of parchment, claiming it to be Caesar's "holy writ": to give away all his fields and assets to the people. This was undoubtedly Antony's killer, the scroll of parchment didn't actually say anything, but public anger had already been aroused. The Roman people began to remember Caesar with sorrow, to sing his praises and to attack in unison the nobles of the Republican party who had taken part in the assassination, which ended in tragedy. The power of language is evident, and the effect of its transmission is very strong, but its drawbacks are also evident in this play.

When human society began, so did the desire to communicate over long distances and the vision of language interchange in other countries.

Return to the biblical story of the Tower of Babel. The Bible (especially the King James Version) is an essential reading for the Western humanities and continues to provide inspiration for Western literature and art today. The allusion to the Tower of Babel naturally became a source of creativity for the filmmakers. In November 2006, the film *The Tower of Heaven* was released in the US and was nominated for several Oscars. The film's narrative is unique in that it places the viewer in God's perspective (also known as omniscient), but presents multiple semi-intelligent perspectives within the film, demonstrating strong dramatic tension and a masterful narration of events. Although the story takes place in modern times, it still draws on the religious metaphor of the Tower of Babel to convey the concern that miscommunication has led to tragedy for human families in various countries. Unlike the Tower of Babel allusion, however, the beginning and the end are slightly reversed: the biblical story of human failure from good communication to poor language and miscommunication; The film *The Tower of Heaven*, on the other hand, is a story of people from different countries coming to terms with each other and understanding each other in the face of language barriers and communication failures that lead to family tragedies. It is clear that, whatever the story, the human desire to share information has always been there. Because it is the key to improving social efficiency and building harmonious interpersonal relationships in society. When language transmission in ancient times no longer met the needs of human society for long-distance information exchange and the breaking of time and space, writing emerged.

Words Create Civilisation

Before the invention of writing, if people needed to keep track of things, they used methods such as knotting. "In ancient times, there was no writing, and the rope was used to record things.", both the ancient Chinese and the Peruvian Indians used to tie ropes to keep track of things, and each had a very systematic and even complex method of recording. The transmission of information by language alone can lead to miscommunication, which is caused by the instability of linguistic information. If we need to pass on certain information from one generation to the next, there must be new vehicles that can take on this task. The advent of text solved this problem by allowing information to be not only shared but also recorded.

The world's earliest writing dates back to around 3000 BC, with the sacred script of ancient Egypt and the cuneiform script of the two river valleys recording the stories of the ancient Egyptian empire, the ancient Babylonian kingdom and the ancient Persian dynasty. These two scripts, which were extinct after B.C., have been revisited by recent archaeologists and are now in museums as important testimonies to the history of writing, and are considered fossils of writing.

It is difficult to determine when the earliest writing in China originated. The Shang dynasty saw the appearance of writing on turtle nails and animal bones, also known as oracle bone inscriptions. In a similar way to the fatalism of the creation of language, the development of writing is also superstitious, but in a very different way from language: shamans played an important role in the spread of writing.

Witches emerged in China in ancient times. In a slave society, where there was a lack of material goods and spirituality, the emergence of witches gave people a certain degree of faith. From a materialist-dialectical point of view, religion, a rather nebulous discipline, was in fact conducive to social harmony if applied properly, as was early witchcraft.

The earliest oracle bones found do not primarily record history, but rather good fortune. By the time of the Yin and Zhou dynasties, witchcraft had become a widespread tool of the ruling class, and the status of the sorcerer rose from that of the common people to that of the ruling nobility. The importance of shamans in the minds of the ruling class in ancient times can still be seen in current films such as the Korean drama *The Moon Embracing the Sun* and the Chinese drama *The Wind Rises in the Forest of Langya*. These sorcerers, although not decisive, were able to gain credibility by virtue of their vast knowledge, their great reputation and the fact that there was no distinction between witchcraft and history: during the Spring and Autumn and Warring States periods, many historians were skilled in divination. As a result of the people's belief in the gods and spirits, the shamans, who were the spokesmen for the gods and spirits, naturally contributed to the development of writing, for as well as divination, they were also responsible for processing the original writing into a full-fledged script and passing it on. The development of witchcraft and the social status of shamans made a very important contribution to the spread of writing.

Over the course of China's long history, through the efforts of countless forefathers, the script has passed through stages such as the oracle bone script, the gold script, the large seal script, the small seal script, the official script, the cursive script, the regular script and the running script. During the Spring and Autumn Period and the Warring States Period, the economic and cultural prosperity led to the widespread use of the Great Seal Script and the Six Kingdoms Old Script as Chinese scripts; after the unification of China by Emperor Qin Shi Huang, the small seal script took over the whole country and was simplified from the previous one. As a means of transmitting information over long distances, writing gave us the power to pass on our understanding of the world, allowing the ancient people to keep abreast of the events of the literary and political worlds: the profound philosophies of Confucius, Mencius and Han Fei during the Spring and Autumn and Warring States periods, as well as the edicts issued by the lord of a country, could be understood by even the illiterate common man through the spread of words and language. At the same time, the words of the people and the lives of the people were passed on through the language into songs, which were later put into writing and transmitted to the ears of the ruling class, thus promoting mutual understanding between the officials and the people.

These songs were first passed down orally among the people and then gradually in written form, resulting in the first collection of poetry in China, the Shi Jing.

The formation of the Poetic Edda spans the period from the early Western Zhou to the Spring and Autumn period, spanning five or six hundred years, and was finally written in the middle of the Spring and Autumn period. The early Poetic Edda is recorded in seal script. This brilliant and subtle Chinese classic is rich in content, ranging from the music and songs of the noble literati to the agricultural work of the common people, and is undoubtedly a treasure of Chinese civilisation. We can read this classic today thanks to the countless Chinese people behind it and the scholars and officials who promoted its transmission, because the transmission of the Psalms was a huge project in the ancient times, when information was not available.

Most of the poems in the Poetic Edda come from the folklore and are difficult to collect alone: it was passed on by word of mouth to a wide range of people, centred on the Yellow River basin, south to the northern bank of the Yangtze River, and then extending to the Jianghan basin. Secondly, these oral ballads took a long time to collect and were slow to be written down: before the invention of the brush and paper, writing was done on paper and silk, which was very labour-intensive and costly. Fortunately, these difficulties were eventually overcome thanks to the hard work and patience of the poetry collectors.

The Psalms could not have been written and passed down without the people behind it.

The first was Yin Jifu, a poet-collector in the Western Zhou, who, in addition to collecting and compiling, was also involved in the composition of some of the chapters of the Psalter. In the Spring and Autumn period, there was a teacher who revised the Psalter and taught it in poems that were widely circulated among his disciples and became very influential, and this teacher was called Kong Zhongni, later known as Confucius.

After Confucius revised and compiled the Poetic Edda, the great task of passing on the tradition was undertaken by one of the disciples, Zixia, one of the "72 sages". In the Western Han Dynasty, the poems of Shen Peigong, Yuan Gu Sheng and Han Ying continued to be handed down, and the poems were called Qi poems, Lu poems and Han poems, collectively known as the "Three Schools of Poetry". The poems of the three families have been successively lost, and only 10 volumes of Han poems remain.

The poems we read today are the "Mao poems" handed down by Lord Mao. On the basis of the previous compilations, the entire book of Mao's poems has a preface, known as the "Great Preface", and a small preface below each poem, which serves to introduce the content and meaning of the piece, known as the "Preface to Mao's Poems". These prefaces were the first Chinese monographs on poetry theory and laid the foundation for the subsequent development of Chinese poetry theory. Although the "Preface to the Poems of Mao" and the "Great Preface" caused some controversy at the time, they are the crystallisation of information that has been passed on and summarised since the invention of writing. It is thanks to these crystallisations that the beginnings of our Confucian poetic tradition were laid, and that our splendid ancient civilisation was written.

The use of text broke the limits of time and space and triggered a second information revolution, compared to the simultaneous transmission of words alone, face to face.

In the late Eastern Han Dynasty, the Chinese script was officially declared to have ended in running script, completing its path of development.

The maturity and widespread use of the written word has brought about an important breakthrough in the recording of information and remote communication. In ancient times, human societies used writing mainly for important matters such as letters and historical records, and with the spread of language, the second information revolution was able to flourish. Early man used a variety of stone and metal tools as writing pens, later on there were quills in the West and brushes in China. Before the advent and immaturity of papermaking, Westerners wrote on parchment and papyrus, while China used paper and silk as writing carriers. Although silk is light, it is not easy to obtain and very expensive, so the ancients mostly used paper documents, mostly bamboo and wooden documents. With the help of the writing on the documents and the diligence of the ancients, much historical information and literature has been preserved, albeit slowly, in the face of poor communication technology. With the help of written records, human civilisation has continued to expand its frontiers, leaving behind many precious documents that have survived to this day.

One of the most worshipped works of all time is the Book of History.

While the historical and literary aspects of The Records of the Grand Historian are well known for their contribution to Chinese history and culture, the process of its production and transmission has been extremely difficult. In the first year of the first century B.C. (104 B.C.), Sima Tan, an official of the Han dynasty who was on his deathbed, left the task of writing history books to his son Sima Qian. As his son inherited his father's work, Sima Qian followed his father's wishes by reading up on history and travelling the country to collect relevant information. The historian's work was marked by many hardships and political torture. It took him more than 10 years to write it and another 10 years to revise it, and it took him a lifetime of work to complete it. Because the book contained unfavourable comments about the Han dynasty, Sima Qian had to carefully hide it from the public after it was written. It was only during the reign of Emperor Xuan of the Han Dynasty that the contents of the book became known to the people, when Sima Qian's grandson, Yang Yun, wrote to the emperor.

The publication of the Historical Records is undoubtedly a milestone in the history of Chinese civilisation, having taken many years to collect the historical materials and over 10 years to write, making a total of over 20 years. In the Western Han period, before paper was available on a large scale, Sima Qian used a brush to write on the paper. The 500,000-word biographical history took a total of 2.4 million bamboo slips. In today's modern world, where books are easily printed and often available electronically, a slightly larger travel backpack or even a very light Kindle can hold this huge volume, whereas in those days it would have filled many wagons and taken up a lot of space in the library. As the earliest editions were made in simple documents, and because of their size, many of the chapters were lost and destruction was inevitable. Although Yang Yun spared no effort to get the book into circulation, it was a slow stream in the Han dynasty, when the means of communication and information were not yet developed.

History is funny: old things are fading away and new things are breaking ground.

When man had a more advanced method of writing than the previous one, he abandoned the backward method of knotting and created human civilisation; as a vehicle for writing, silk and silk receded from history with the advent of a more advanced technology, the printing press.

The Printing Press Drives Ancient Civilisation

The desire to transmit information over long distances has always existed, even in ancient times when the need was not so great. For this reason, the wise people of ancient times devised many methods of long-distance transmission, such as wolf smoke, beacons, stagecoaches and carrier pigeons.

During the Western Zhou period, the method of transmitting information by beacon was well established: beacons were built along the route from the capital to the border for the transmission of military information. When the enemy invades, the sentry lights a beacon to be seen by the surrounding garrisons, who then light the beacon in turn, transmitting the information long-distance and sending reinforcements to the front line in time to support them. There is no doubt that it was a military stronghold for the country at the time, a matter of life and death.

Through the clouds of history, the story of the King of Zhou and his lords has been passed down to this day. The King of Zhou, whose favourite concubine was not fond of smiling, listened to the slanderous words of the treacherous Guo Shifu in order to make her laugh: light the beacon. When the lords saw this, they thought that the emperor was in a state of emergency and quickly came to the rescue with heavy troops. When they arrived, there were no enemy troops in sight, only King Zhou and Qiaosi drinking and having fun. The audience was in a state of panic and when they saw this scene, they could not help but laugh. The king was so pleased that he used the same method to please her several times afterwards. The story of the wolf came to life, and when the enemy really did take the capital, Haojing, and the beacon was lit, no one came back, King You of Zhou was killed, Bao Si was captured, and the Western Zhou was destroyed. This is the most classic story of ancient telematics. By the time of printing, human society had achieved not only long-distance transmission, but also the mass transmission of information.

The invention of paper was essential to the creation of the printing press. In the process of using the cumbersome and cumbersome abbreviations, people were constantly searching for new text carriers to convey information more conveniently. The earliest paper was produced in rags, made from old fishing nets and twine ends, this fibre paper is too crude to be used for writing. In the middle of the Eastern Han Dynasty (105 AD), Cai Lun used the bark of a tree to improve the original paper, making the famous "Caihou Paper", which he taught and spread from Henan to the rest of the world.

From the end of the Eastern Han Dynasty to the Three Kingdoms and the Jin Dynasty, history has brought with it a golden age of cultural development. The Spring and Autumn period and the Warring States period, when a hundred schools of philosophical thought competed; during the two Jin dynasties, there were many famous painters and calligraphers. Wang Xizhi, the calligrapher of the Eastern Jin dynasty, was the representative of this group of works, which led to the development of paper for painting and calligraphy. At this time hemp and kozo bark were added to the raw material for the paper, improving the quality of the writing paper.

During the Tang and Song dynasties, poetry flourished, bamboo paper emerged and became very popular among the literati, and paper types diversified: in addition to bamboo paper, there is also linen paper, leather paper and rattan paper, and the techniques used to make them are no longer homogeneous. The choice of raw materials has also been extended to include bamboo pulp, rice straw and wheat straw.

Thanks to the invention of paper, the cost of recording information was greatly reduced, and the speed of dissemination of works such as the Histories was finally increased, allowing people to hand-copy books and pass them on.

In the middle of the Qing Dynasty, China's hand-made paper technology was already very advanced, and the high quality and variety of paper, combined with the desire for information to travel further, provided a solid backbone for the third information revolution.

History has evolved, and as time went on, people demanded more information and the printing press came into being.

Inspired by the carving of seals, the earliest use of engraving was in China, where hard wood was used to carve out the reverse characters, which were then inked and painted. The books produced by this technique are so beautiful that the Vajra Paramita Sutra engravings that have survived can still be seen in modern times, and some museums offer hands-on engraving and printing experiences for visitors to appreciate the mastery of ancient Chinese techniques. The engraving and printing of books greatly accelerated the production of finished books and increased the efficiency of information transmission compared to hand-copied books. However, the drawbacks are also obvious: each book must be engraved with a set of printing plates, high cost, occupy a lot of space, and printing colours are easy to mix, resulting in clear boundaries of colour blocks. To solve the problems caused by engraving, movable type was invented.

During the Qingli period of the Northern Song Dynasty (1041–1048), the civilian inventor Bi Sheng invented movable type by using clay. As one of the four great inventions of China, movable type solved the drawbacks of engraving and could be used repeatedly, with more flexible typesetting than before. Based on the technical ideas of movable type printing, Wang Zhen replaced clay movable type with wood movable type in the Yuan Dynasty, and invented the rotating type frame, followed by tin movable type and lead movable type. With the creation of movable type, the information revolution took another major turn. Thanks to the cultural exchanges between China and Europe during the Song and Yuan dynasties, movable type printing gradually spread to Europe, leading to a worldwide change in information transmission technology, an unprecedented civilisation and the formalisation of the traditional publishing system.

The printing press originated in China, but its quantified use was in the West.

When hemp paper technology from the Tang Dynasty was introduced to the West, the costly parchment was gradually replaced by cheaper hemp paper. It is saddening to note that while China was moving towards seclusion, the industrial revolution in the West was taking hold and paper production was moving from hand-made to machine-made. When printing spread to the West, Westerners began to find ways to quantify it. The invention of lead type by the German Gutenberg in 1455, followed by the successful implementation of machine printing technology, led to a qualitative breakthrough in the efficiency of information transmission. As the saying goes, the use of printing machines in large numbers coincided with the European Renaissance. Since then, European economic, technological, religious, literary and artistic developments have exploded, rapidly sweeping through Western capitalist society and making a huge contribution to enriching the treasures of human civilisation worldwide.

The development of the printing press made it possible to transmit information over large distances and to mass-produce books with their intellectual content, which flowed rapidly into society. The fall in costs made it possible for ordinary people to acquire knowledge and ideas from books. In Western societies, where there was no television or even electricity, people's access to information came mainly from books, newspapers and letters, and they were influenced in their daily lives by the content of the information contained in books.

As the British theatre became the mainstream of literature, dramatic writers proliferated and plays were edited and published for public consumption. After the dark years of the Middle Ages and the obscure years of the Stuart Restoration, people in the midst of the chaos of the world read books for their own amusement, entertained themselves in family theatrical performances, or entertained their friends and relatives by reading them aloud, so that they could learn about the political and social conditions of the time. The sophistication of the printing and publishing industry has allowed information to be successfully transmitted over long distances, breaking down geographical limitations and allowing people to read and absorb foreign works, thus contributing to the development of their own culture. Chaucer, the father of English literature, was inspired by the Italian poet Dante's Divine Comedy to create The Council of the Hundred Birds, and by Boccaccio's Decameron to create the classic Canterbury Tales, which have had a profound impact to this day. At the height of the novel's popularity in the eighteenth century, many famous writers wrote novels about the social phenomena of their time, enlightening the public with their ideas and fulfilling their mission of mass communication.

As an important vehicle for the third information revolution, books have not only influenced people's lifestyles, but have also contributed to the development of new ideas. In the mid-nineteenth century, the famous English writer Charlotte Brontë shocked the literary world with the publication of her masterpiece *Jane Eyre*. By this time, Britain had become the Empire of the Sun after the reign of Elizabeth I, but the status of women was still very low: few jobs were available to women, and women had no inheritance rights. *Jane Eyre*, on the other hand, ends with the heroine inheriting a large estate and taking charge of Mr Rochester's estate, which was daring, even treacherous, for the time. The novel was a huge sensation, marking the beginning of feminism, and since then women have emerged as writers who communicate the awakening of female consciousness through their pens. Looking back on history, one cannot help but be amazed at the profound impact that the information revolution has had on the spiritual fabric of human society, and the development of the printing press has made it possible for once great works to be brought back to life.

One of the most iconic examples of this is the poetry of the American Imagist poet Emily Dickinson.

Dickinson lived a lonely life, and despite her family's prominence and wealth, she was misunderstood and isolated from the world from the age of 25. After Dickinson's death, her sister Lavinia stumbled upon a large number of manuscripts that she had left behind, with over a thousand poems in the manuscript, I realised how passionate my sister was about literature during her lifetime. Lavinia worked on all fronts to get these poems published, finally, with the support of his editor Todd, Dickinson's poems were published in 1890. It was only after the poems were published that the literary world recognised the creative genius of this talented poet. The freshness of the imagery, the depth of thought and the unique style of Dickinson's poems have touched the hearts of the world and have been met with critical acclaim.

It is no coincidence that Sima Qian, the author of The Records of the Grand Historian, did not see his work published during his lifetime, centuries after it was written and published in an era of poor information transmission; in contrast, Dickinson's poems were widely circulated only four years after his death, aided by advanced printing technology. Without the boom in papermaking and printing, Dickinson's poems would have been permanently buried, passed around among families, and Western literature would have lost one of the world's great writers on a par with Shakespeare and Lev Tolstoy.

The third information revolution is undoubtedly one of the most significant in the history of the information revolution, with people communicating over long distances through writing and paper, reading books to learn about social events, having a wide range of entertainment and idolatry. To this day, in the deep nostalgia of the British people, the period around the eighteenth century is the one to which they most aspire. At a time when information delivery was exploding and people's growing needs were giving rise to faster means of communication, when people were impatiently waiting for a reply and anxiously worrying about whether their letters had been lost in transit, radio came into the picture.

Radio Leads Modern Civilisation

As a legacy of the third information revolution, the use of books and paper to convey information is still widely used today. However, there is a problem of delay in both the mailing of letters and the publication of books: it can take days, even months, from the time the letter is sent to the time it is received; it takes longer to write, revise, finalise and publish a book, the problem of this delay is becoming increasingly difficult to meet the urgent need for information. The advent of radio has made it possible to transmit information over long distances and in real time, allowing human society to leave the traditional way of life behind and to ascend the ladder of modern history.

Before we talk about radio communication, we must first talk about the creation of electricity.

Electricity has been explored since ancient times, but almost always in the direction of gods and ghosts. The real scientific exploration of electricity began in Western society around the eighteenth century with the famous American scientist and statesman Franklin. Based on his study of lightning in the sky, Franklin developed the concept of "electric current" to explore whether electricity of the same nature as lightning also existed on the ground. His widely known kite experiments gave full credence to this claim and led to the invention of the world's first lightning rod, based on the principle of metal conductor diversion in his experiments.

The kite experiment caused a great sensation in the Western world, confirming the existence of electric currents and bringing light to further exploration. On a lightning day in 1780, an Italian professor of anatomy, Galvani, happened to see a spasm in the leg of a frog on a dissecting table and, filled with curiosity, began more than a decade of research, which revealed that the phenomenon was an electric circuit and that he believed that the spasm occurred in the frog because of the presence of electricity in the animal. In 1791, Galvani published the results of his research on spasticity of the frog's legs, which caused a scientific sensation. However, another Italian scientist, Volt, questioned Galvani's view, he believed that electricity was not found in the muscles of animals, but in metals. In 1799, Volt experimented to create the world's first battery, the Volt battery, the paper in question was published a year later at the Royal Society.

A major breakthrough in electricity also came at this time.

Born on 22 September 1791 into a poor blacksmith's family in Surrey, England, the boy spent his childhood in the grip of hunger and cold and was taught by his father to be resolute and industrious. Too poor to attend school, the boy was forced to work as a newspaper boy at the age of 12, and then as an apprentice to a stapler. Although he did not receive a formal education, thanks to the popularity of books in the third information revolution, he saw mountains of books in the stapler's house and, out of a strong desire for knowledge, he began to read with a thirst and accumulated a lot of knowledge of natural sciences on his own. His attention was drawn to electricity in the Encyclopaedia Britannica and he used its knowledge to teach himself to experiment with waste and to set up study groups with friends and become active in scientific organisations. The boy's hard work and learning brought him opportunities: on the recommendation of a bookshop customer, he listened to a lecture by the famous chemist David, and took private notes for David. At the age of 20, he became David's laboratory assistant.

This boy was Faraday, the father of electricity in the world.

In 1831, Faraday discovered the phenomenon of electromagnetic induction and experimented with a method for generating alternating current, a groundbreaking discovery that has made a huge contribution to human civilisation. Faraday then soon invented the disc dynamo, the world's first generator.

In 1852, Faraday introduced the concept of magnetic lines of force, laying the theoretical foundations of electromagnetism. Based on Faraday's work, the British physicist Maxwell developed the theory of classical electromagnetism. Scientific exploration is always on the shoulders of giants, as the German physicist Hertz did in his scientific career, experimental proof of Maxwell's theory of electromagnetism, confirmation of the existence of electromagnetic waves and rewriting of Maxwell's equations. Through his experiments, Hertz proved that electrical signals can travel through air and also discovered the photoelectric effect through the phenomenon of ultraviolet light shining on electrically charged objects.

The maturity of electromagnetism theory laid a solid foundation for the creation of radio.

There is much debate in academic circles as to who started and invented radio, and the technological rivalry between the great powers began with radio.

The British believe that Maxwell pioneered radio because he was the first to suggest the existence of electromagnetic waves; the Russians, however, only recognise the Popovs of their country; the Germans consider Hertz to be the inventor of radio because he was the first to prove the existence of electromagnetic waves; anyone who knows Tesla recognises his status as the father of radio; but in the eyes of Italian scientists, Marconi invented radio communication and won the Nobel Prize in Physics.

The emergence of radio is the inevitable product of the world's scientific and technological progress, not the originality of any one person, but the result of the joint efforts of scientists, and its application has contributed to the development of modern civilisation in human society.

In 1837, the Englishmen Charles Wheatstone and William Cook patented the telegraph lines they had developed.

In 1893, the Serbian-American scientist Tesla demonstrated radio communication for the first time in St. Louis, Missouri. In 1894, the Russian scientist Popov proved that "electromagnetic waves can be used to send signals to distant places" and invented the world's first radio receiver. In the same year, at the age of 20, Marconi began experiments on the long-distance transmission of electromagnetic waves and in 1895 succeeded in extending the transmission distance to 2.7 km; In 1896 he ran to England with his rudimentary launcher and again extended his experiments to a distance of about 14.4 km; In July 1897, the Marconi Radio Telegraph and Signal Company Ltd. was founded.

In 1901, Marconi successfully completed 3600 kilometres of radio communication across the Atlantic Ocean, before his wireless telegraphy was commercially available; at the same time, Popov put radio into military service and set up a wireless communication network of more than 40 kilometres.

The fourth information revolution is announced.

Who is the true father of radio? To this day, the debate never ends, but the scientists' remarkable contribution to the progress of human society will always be revered and remembered.

From the telegraph came the telephone and radio.

At the beginning of the twentieth century, audio broadcasting was introduced, firstly as marine radio telegraphy, using Morse code. The code was a signal code used in early radio communications and was so clear and simple that it was also commonly used for underground intelligence work in warfare, as illustrated in the Chinese film *The Sound of the Wind*.

At a time when Marconi's long-distance radio experiments were achieving remarkable results, China was no longer in the heyday of the

Spring and Autumn Period and the Warring States, and the cultural heyday of papermaking and printing was in decline. At the end of the Qing dynasty, the world was in dire straits and the ruling class was on the rocks.

The Qing dynasty, which had adopted a closed-door approach and rejected modern science and technology, began to pay a terrible price in 1840 when it was attacked by Britain with advanced heat weapons. At the end of the Qing Dynasty, under the influence of the foreign affairs movement and other factors, a Chinese translation of the newspaper "Radio" was published in Book 25 of the "Times", which was founded by Wang Kangnian, Liang Qichao and other leaders of the Restoration Movement, and this event was the beginning of the introduction of radio communication technology into China.

As the wheels of history turned into the twentieth century, radio communications took over and the days of horse-drawn carriages and messengers as information hubs no longer existed.

The outbreak of the First World War plunged the world into misery, while radio was used in large numbers for military purposes and saw its own success. The commanders of the warring nations used radio as an important channel for the transmission of information, enabling them to keep abreast of the situation quickly. By the Second World War, some 60,000 people in the United States had radio licences, about 90% of which served the war effort and the military industry. And when Nazi Germany wanted to occupy a country, the first means of doing so was to take control of the radio. As an end product of radio technology, radio stations and radios have also been a major tool of the Communist underground in China, and remain an important tool of national security and stability today. After the founding of the People's Republic of China, radio became fully popular in an era of peaceful development.

Broadcasting is an important vehicle for radio.

The Shawshank Redemption, by American author Stephen King, was brought to the big screen in 1994 and would have won all the major Oscars had it not been for Forrest Gump at the same time. Although it lost out to the Oscars, the film was an overwhelming word-of-mouth triumph. There are several classic scenes in the film, one of which is this: Andy finds records and a gramophone in the warden's office, along with the accompanying radio equipment, and improvises by playing the records without permission, locking the door and turning on the radio. It was lunchtime and all the inmates were out on the playground when the tannoy in the playground blared music and everyone looked up and stopped as the third act of Mozart's opera "The Marriage of Figaro" was sung over the radio and echoed throughout Shawshank Prison.

Radio broadcasting as the main means of transmitting information about the war, as illustrated by the Oscar-winning film *The King's Speech*: George VI was unable to speak in public because of his stammer, but with the help of his speech therapist Logue, he was able to overcome this obstacle and deliver an inspiring speech before the Second World War. The fourth information revolution has enabled the simultaneous remote transmission of voice, breaking through the limitations of text, distance and latency, allowing people to directly access and empathise with instant information content through sound. Andy plays opera music to give the inmates a moment of freedom; In war-time Britain, George VI's inspiring speeches via radio technology shook people to their core. The advantages of mass communication offered by radio are evident here.

The printing press made it possible to transmit texts by means of books; Radio broadcasting has made radio possible; The telephone and radio allow voice to be transmitted in real time. Information and communication has undergone a transformation and, after the transmission of text and voice, will move towards the transmission of images and video.

Next, the mass media came to the fore, taking information dissemination to new heights.

Television Advances Modern Civilisation

Every technological innovation is accompanied by controversy, and information transfer technology, one of the hallmarks of human civilisation, is at the heart of this controversy. On 11 November 2016, internationally acclaimed Chinese director Ang Lee's new film *Billy Lynn's Battle in the Middle* was released in China, using 120 fps filming technology for the first time, providing a more immersive experience than ordinary 3D films, but also inevitably leading to much controversy. The video image transmission technology of television is similar to that of film, using the visual residual effect of the human eye to show a frame by frame fading still image, forming a dynamic picture.

If the inventor of radio was controversial, the invention of television was accompanied by the tears of the people behind it.

Television first appeared in Britain in 1925, when a man called Baird built a mechanical television set. The material used in the production of this TV is almost scrap: optical equipment made from bicycle lights and enamel pots to build the frame. However, the machine, which is shaped like a black box, reveals faint but lifelike images of puppets. Baird worked on the development of television sets using mechanical scanning technology and developed the first colour television set in 1928. Inspired by the radio, he ventured a hypothesis: as electricity can be used to transmit voice, it can also be used to transmit images. However, just as Baird was attracting investors' attention, the American inventor Farnsworth's electronically produced television set took down the mechanical technology and soon took over the market, leaving Baird to die with regret.

Farnsworth was a boy genius from an early age, and at the age of 14 he had a very different intuition when Baird was working on a mechanical television set: mechanical technology cannot transmit images, electronic technology can, because machinery can never operate at a speed that would allow it to capture electronic signals. According to Farnsworth's reasoning, the picture, if converted into a stream of electrons, can be transmuted through space like a radio wave and finally aggregated into an image by the receiving terminal.

After graduating from high school, Farnsworth attended Brigham Young University, but had to interrupt his studies due to family changes. He moved to San Francisco to work hard on a television set, and in 1927 the first working television receiver and picture tube were developed. When Farnsworth's efforts finally caught the attention of investors and the government granted him a patent, the American Broadcasting Corporation jumped in to fight Farnsworth for the rights to the invention, arguing that Farnsworth's idea of development at the tender age of 14 was a lie and using the excuse that he had not yet graduated from university to argue that Farnsworth could not possibly have the ability to invent a television set. Although Farnsworth produced strong witnesses and evidence to win the case for himself, he no longer had the funds to promote his invention.

Worse is yet to come.

In 1930, a visitor to the Farnsworth Television Laboratory in San Francisco, claiming to be a television enthusiast, visited Farnsworth and asked him for advice, and spent three days in the laboratory. Three years later, the American Radio Company produced a television set and made a big splash. It was then that Farnsworth realised that the guest who had visited the company three years earlier was Zorikin, a television inventor employed by the company, who had built a prototype at the time, but the finished product was not working well and had come to steal the core technology by concealing his identity in order to obtain it. At the time, the major electronics companies had already signed patent contracts with Radio America and were therefore afraid to invest in Farnsworth. Thus began another protracted patent battle. The case was long and stinky, and Radio America lost and then appealed, dragging on for years.

When Farnsworth finally legally patented the television invention, he was almost penniless, and with the Second World War about to begin, the US government was forced to suspend the television industry. By the time it could be produced again, Farnsworth's patent had expired. This was the moment when the American Radio Company began mass production of televisions and made a big deal out of it, making Zollicin, who had stolen Farnsworth's secret technology, the father of television. As desperate and powerless as an old king watching a Trojan horse attack his city, Farnsworth was disillusioned and returned to his homeland, where he fell ill.

Although the inception of television was a sad one from the start, this did not stop it from sweeping the world.

On 2 November 1936, the day the world of television was born, the BBC launched a television programme. At the 1939 World's Fair in New York, the television set made a splash and became popular after the war, adding to the modern civilisation of human society.

The first colour television set was introduced by Radio Corporation of America in 1954, but the picture quality was so poor and the cost of installation so high that many households opted for black and white television. The 1970s saw the opening of a diversified route for the transmission of information via television equipment, as the development of multi-channel television equipment meant that people could see a wide range of television programmes to meet the needs of different audiences.

The spread of television marked the birth of multimedia, combining sound, text, images and video, allowing information to be transmitted in real time, on a large scale and over long distances. What is more important is that the public has a visual sense of what is going on, and the message as a vehicle has since taken on an emotional dimension.

Such sentimentality was typical of the coverage of the Vietnam War.

In the early years of the war, influenced by heroism and romanticism, the vast majority of American soldiers were brave and courageous, and the public was treated to glowing images of the heroes of the war. As time went on, the situation reversed. Especially towards the end of the war, there was a strong anti-war wave in the United States: Many peace groups held mass demonstrations, university students joined the anti-war movement, and pop singers released anti-war songs. Much of the reason for this dramatic shift in the American people's attitude to the war is due to television coverage. When people see the bloody scenes of war and brutal shootings on television, they are visually shocked. The impact of television media is completely different from that of newspapers and radio: Newspaper and radio reports of the battle and the number of casualties do not have a strong sense of immersion for the general public, but television presents the images directly to the public, creating a visceral and powerful emotion. The public finally realised, through the brutal images shown on television, that the scenes in *Farewell to Arms* and *The Sun Also Rises* were the real thing, that there was no such thing as heroism or romance on the battlefield, only naked bloodshed and despair.

As an important vehicle for multimedia, the television allows for a richer, more emotional and impactful way of delivering information. It became one of the icons of modern civilisation and, as television became more popular in China, it became a standard feature in the homes of ordinary people.

In March 1958, China's first television set was produced, developed by the state-owned Tianjin Radio Factory. In the 1980s, the television set accelerated and became the dominant force in the dissemination of information. Compared to radio broadcasts, television not only received all kinds of social news, but also had sound and images, giving the people a more visual experience, and both entertainment functions, and became a rare item upon its introduction. At that time, whenever anyone had a television set, in the evening, all the neighbours who did not have a television set would gather at his house to watch television programmes. Older TV brands include "Feiyue", "Golden Star", "Peony" and many others. Owning a television is seen as a symbol of wealth. On the streets of Chinese cities, whenever there is a television set on display in the window, people are bound to gather to watch it. Later, with the introduction of colour television, a number of foreign television brands entered China, with Japan's "Toshiba" being the most widely known.

In 1994, Beijing TV launched Animal Paradise, a large-scale variety programme that combines knowledge, entertainment and fun, which was broadcast by Toshiba Corporation of Japan as a special title. The programme was an instant hit and ran for 12 years, with record ratings, especially during the period when the famous host Wang Gang was the owner. The programme not only opened the eyes of people living in the city by giving them comprehensive information about the various animals in nature, but also made Toshiba TV sets a hit in China. At that time, the price of a Toshiba television set was over $3000, which was equivalent to one to two years of an average person's salary, but this did not stop Toshiba televisions from selling well.

In contrast to traditional paper and radio broadcasting, the development of television programmes has brought a wide range of entertainment and entertaining knowledge, and people's access to information is no longer limited to political speeches, war reports and current affairs. With the spread of television sets, major television stations have been established and are actively developing their own television programmes and developing the information and culture industry using television as a vehicle. In September 1958, after meticulous preparations, Beijing Television (renamed China Central Television in May 1978) made its first official broadcast of television programmes, and cultural construction began to develop steadily. Although Beijing TV was forced to stop broadcasting during the 10 years of turmoil, the official launch of "News Broadcast" in 1978 started the prelude to CCTV. After decades of development, CCTV has become the largest news and cultural communication forum in China, carrying the important mission of transmitting information at home and abroad.

As a vehicle for the fifth information revolution, television is no longer limited to news and current affairs, but also includes a wide range of television programmes of an entertaining nature, and the television drama industry has emerged.

In the 1990s, when the TV series *Desire* was broadcast, it was a hit with millions of people. The drama's portrayal of the emotional lives of small characters against the backdrop of a decade of turmoil resonates deeply with viewers and demonstrates the true power of television as a message delivery tool. Since then, *Beijingers in New York* has been aired and has quickly become a hit. At that time, the American people had no idea about China, only that it was a distant and unfamiliar country, and the fate of the protagonists in the play in New York touched the hearts of millions of people. The scenes are shown, the characters are spoken, and the audience is given a very visual perception of the message on the other side of the ocean, an effect that radio cannot achieve. In the radio era, information was inevitably lost in the output process, while products with television as the information terminal greatly improved the integrity of information transmission. People also have a much more comfortable experience than radio when receiving information.

The rise and popularity of television has advanced the modern civilisation of human society, but over time it has become unable to meet the growing individual needs of people. When society is relatively well-off in material terms and peace and development are the themes of the times, the spiritual needs of the general public and the desire to share information grows. The great convergence of world civilisations and the latest trends in economic globalisation have once again placed greater demands on the information revolution. This time, the more powerful Internet was on the scene.

There has never been a more exciting time than the dawn of the Internet age.

The Internet Ignites Contemporary Civilisation

Television has allowed multimedia to emerge strongly, but its information transmission is one-way, we can only receive the given information, but cannot transmit our own information, two-way communication cannot rely on television to achieve. The emergence of the Internet has brought information transmission to the highest level in the history of the information revolution, effectively combining all the characteristics of previous information carriers: Real-time, long-distance and multimedia, with the added advantage of two-way information interchange.

When the Soviet Union launched Sputnik, the first artificial Earth satellite, in October 1957, the US government panicked and rushed the Department of Defense to form the Advanced Research Projects Agency (ARPA). In December 1969, ARPA brought together UCLA and Santa Barbara, Four mainframes at the Stanford Research Institute and the University of Utah were connected to create the Arpanet, or ApNet, a network used by the military. This was the earliest network, using packet-switching technology and operating at only 50Kbps. A year later, the network working group developed a communication protocol between the hosts to control the transmission of signals over the network. After another year, communication software was developed to enable communication between hosts, and this was email. The birth of email made Internet communication immediately efficient and fast. At this point, the Apa network began to grow in size, leading to the widely known TCP/IP protocol.

By the early 1980s, computer networks had become very diverse, including USENET, CSNET and others. The advent of the NSF, a wide area network funded by the US National Science Foundation, gave a major boost to the development of the Internet. In the early days the Internet was used for research in universities, and as the number of hosts increased dramatically, more and more people used the Internet as a tool for information exchange. In the 1990s, with the advent of browsers and web technologies, the Internet saw rapid growth, and in 1995, the NSF network was launched commercially and the Internet began to sweep the world.

The sixth information revolution is ready to explode onto the market, bringing a great baptism to human society.

In 1994, the Chinese government supported the construction of the CERNET demonstration network project, which established the first TCP/IP Internet in China. Zhang Shuxin founded the first Internet service provider in China, Yinghaiwei, and the Internet began to reach millions of households, and soon entered a phase of rapid development. From the beginning of the four major portals (Netease, Sohu, Tencent and Sina), to the current boom in multimedia and e-commerce, people are already familiar with the identity of Internet users. Compared to television transmission, the Internet carries a more complete message, and contemporary human civilisation has been made brighter by the information storm it has unleashed.

The year 1998 was a major turning point in the development of the Internet in China. In June of this year, a generation of classic Microsoft Windows 98 operating system was released, IDSN became popular in China, and BBS forums and online chat ushered in a golden period of development. The changes brought about by the Internet are also driving huge changes in the entertainment and cultural industries. It was also the year that the first online novel, First Close Encounters, entered the limelight. The novel is based on the main story of Internet chat, The plot seems rather old-fashioned now, but it was a big hit at the time. This innocent, mischievous and tragic novel of online romance was the beginning of online literature and the prologue to popular online literature. Since then, the tastes of Internet readers have risen and good works have been produced on the Internet, giving rise to a large number of famous Internet writers, such as Nanpai Sanshu, Tang Jia Sanshou and Liu Tu Zi, who have made a name for themselves on the Internet and garnered numerous fans.

Many of those who were at the forefront of the information revolution have benefited and even gone down in history. During the first wave of the Internet, many of those who fought the waves on the front line became outstanding representatives of contemporary culture, and some even changed the course of society and people's habits significantly.

With the start of the Internet in China, a group of science and engineering students smelled a huge business opportunity and began to work quietly.

In 1994, Ding Lei, who was working in the Ningbo Telecom Bureau, felt the joy of the Internet when he first touched it. A year later, despite strong opposition from his family, he resigned from the telecoms bureau and started to venture into the world. In May 1997, NetEase was founded. In May 1997, NetEase was founded, and surprisingly, in just three years, in June 2000, NetEase was successfully listed on NASDAQ.

The other student has had a smooth ride compared to Ding Lei's career ups and downs. In 1996, Zhang Chaoyang, a post-doctoral fellow from MIT who had returned to China with venture capital, founded AIT, and two years later, the company launched its own portal, which was renamed Sohu. Zhang Chaoyang's good educational background and superb execution allowed Sohu to grow extremely rapidly. In July 2000, Sohu.com was listed on Nasdaq.

Unlike these two young talents who started their own business within a few years of leaving campus, Wang Zhidong, who graduated from Peking University in 1988, worked in the field of software development and management for 10 years, gaining a lot of experience before starting his own business. In 1998, Sina was founded and grew rapidly under the leadership of Wang Zhidong, and was listed on the NASDAQ in the same year as Sohu.

Around that time, QQ came along. In 1999, OICQ was launched, focusing on live online chat, a way of typing and communicating with strangers remotely, which quickly gained a foothold in the Chinese market. Despite the lawsuit over its early name, the renamed QQ brought huge profits to Internet cafes in the streets, and Tencent went on to develop games, space, farms and other entertainment applications. Despite the social problems and negative press caused by online chatting, this has not affected Tencent's strides. In June 2004, Tencent was listed on the Main Board of the Hong Kong Stock Exchange.

The four major Internet portals in China once formed a four-legged triumph, and as the first echelon of Internet companies, they are still standing.

After 2000, online games emerged and blogs became popular. In the new industry of the Internet, Ma Huateng, Ding Lei, Zhang Chaoyang and others seized the opportunity of the first wave of information and let the public experience the changes of the times.

On 12 May 2008, a massive earthquake measuring 8.0 on the Richter scale struck Wenchuan, Sichuan, which soon made headlines in the media at home and abroad. The Chinese government and the general public were able to act quickly, accurately and efficiently in this disaster, influenced in large part by the Internet age.

Within 25 minutes of the quake, Sina.com posted an accurate news report, and all three major portals soon launched special features on the quake to report on its progress, and the news soon spread throughout the country. Within a few hours, videos, blogs and online media reports were updated. The central government also promptly issued policies and implemented emergency relief operations.

The rapid dissemination of information has had a profound impact on all aspects: Although communications in the disaster area were once paralysed and contact with the outside world was lost, major operators soon opened free dialing services for communications in the disaster area, and information links between the disaster area and the outside world were gradually restored. With more ways of transmitting information than ever before, and with the media constantly following the event in real time, the world was able to see not only the speed of the Chinese government's decisions and the strength of its people, but also how fast the latest methods of transmitting information can be.

All of this would have been unthinkable in a time when information was not available. Traditional media and new media have been strongly influenced by the Internet, allowing information to be delivered as fast as possible, and the delay in people receiving news has been greatly reduced.

These are bad times, and the information explosion is causing a lot of social problems: Internet addiction, cybercrime, fake news, etc.; these are also wonderful times, when all kinds of information are efficiently shared, social benefits are greatly enhanced, and transparency is greatly increased …

History is always progressing, and the four major portals are inevitably facing transformation in the mobile Internet era, while Tencent has emerged as a leader once again with the development of WeChat.

When the Internet shifted from PC to mobile and the role of mobile phones began to outweigh that of computers, a new era was about to emerge.

This time, the more powerful Smart Internet is no longer a science fiction scenario from the movies, but is gradually entering people's lives and will continue to change the world.

Smart Internet and the Seventh Information Revolution

In the history of the six major information revolutions that preceded it, we have seen a lineage of information delivery and development that raises a very important question as the Internet fuels the leapfrogging of the information revolution: What will the seventh information revolution be? The answer is already obvious: the Smart Internet.

However, it is important to stress that: The Smart Internet is more than just the Internet.

Previously, the Internet was referred to as the traditional Internet, and its main task was to enable the transmission of information at high speeds. The earliest ApNet was a network of servers connected by the TC protocol, and in the commercial era the fundamental problem solved by the Internet was to make the transmission of information unobstructed and to convey the basic spirit of the Internet: Freedom, openness and sharing.

However, the Smart Internet is no longer just the Internet, but a new information delivery system based on the traditional Internet, built from mobile connectivity, intelligent sensing, big data and intelligent learning, it is more comprehensive and powerful, with more obvious intelligent features.

Mobile Connectivity Is the Basis for Intelligence

Traditionally, the Internet has been used to establish information links between each computer, each computer has its own fixed IP address and is connected by cables and fibre optic cables with a fixed termination system.

When the mobile phone became Internet enabled, various related applications were gradually expanded from the PC side to the mobile side, from the initial use of chat applications and web browsers on the mobile phone to almost all the applications that users need today. Mobile Internet is mobile, mobile is the core and foundation of the entire mobile Internet, its terminal is no longer just a computer, but the terminal of all mobile products, first there are mobile phones, and then there are other products, even cars.

Unlike the old days of computer access, a mobile phone is now all that is needed to break the space barrier and connect to the Internet at any time, so that any information you want is just a few keystrokes away. In addition to mobile phones, iPads and smart bracelets are also available as end products for mobile communication. With the Internet booming in human society, the computer is no longer the only Internet terminal, mobile connectivity is the main theme of the world today and will have a domino effect, driving the unpredictable but traceable information age forward.

As a result, the mobile Internet is expanding the use of scenarios exponentially and breaking the limits of space, while 5G is the foundation of mobile Internet.

As an important carrier of the current information revolution, the Internet will evolve into a smart Internet, triggering a new round of information revolution towards total intelligence.

In today's era of information explosion, the smartphone has been popularised from its introduction to the mobile communication industry and has laid a solid foundation for the future smart era. The mobile phone is no longer just a mobile phone for talking and texting, but has become a smart terminal with many functions: The phone is only the most basic function, but the real purpose is to transmit information on everything, to make mobile payments, and to combine various multimedia applications to become a universal machine that is essential for people's work and life.

Mobile Internet plays a strong role in supporting the construction of smart Internet. The concept of mobile payment was introduced a long time ago, but this function could not be developed in the 2G and 3G era, mainly because the network communication capability was not strong enough, and the payment function had a long delay and was slow to respond. In the 4G era, mobile payments in China have exploded because of the unprecedented strength of mobile network signals and the extensive coverage of 4G networks. And 5G must provide strong support for the services of the Smart Internet, Without 5G, the role of the Smart Internet will not be fulfilled and the application of artificial intelligence will not be possible.

With the popularity of smartphones, the smart Internet is beginning to emerge and various mobile smart products are beginning to influence various industries and will gradually realise the interconnection of people and things and the interconnection of everything. At the moment, the Internet of Things is taking shape, but it is not yet in full bloom, this is because the 5G network has not yet been truly established, and once the 5G era arrives, the mobile Internet will certainly set off a new round of information revolution.

The most intuitive sense of this storm is that everything is connected, and that the Internet of Everything relies on intelligent sensing to take root.

Intelligent Sensing Extends the Human Organ

On 30 March 2018, the famous director Spielberg's masterpiece *Top Gun* was released in China and exploded the entire movie market. Fans were amazed by the numerous hidden eggs in the film for anime fans, movie fans and game fans. The heavy industrial feel of the future world created in the film scenes, the seamless integration of virtual reality, all the characters wearing "Oasis" game sensors and getting a realistic touch in the virtual world, and the visual impact of the cool sci-fi experience … made people addicted to it and their two eyes were completely inadequate.

It's clear that the Hollywood filmmaker's unconventional yet logical imagination gives us a real sense of the saying "nothing is invisible, only unthinkable", with the main line of characters wearing sensors from start to finish and drones searching for the player's geographic location throughout the film, the use of smart sensors in the future reaching its peak.

Smart sensing is the cornerstone of the success of the Internet of Things. With the development of information dissemination technology and the 4G network, the future 5G era will provide an intuitive experience for users with the booming development of smart sensing. And that fire is now going strong, although it's still a long way from the heights of Top Gun, however, numerous sensors are now accepted on the market. These sensors are recording the world in a cellular fashion, forming new systems.

From the point of view of "induction" alone, its power is already omnipresent, a smartphone alone is equipped with gravity sensing, pressure sensing, touch sensing, radiation sensing, image sensing, face recognition and many other functions, as well as being able to sense the outside world through various intelligent recognitions.

The most representative of these is geolocation sensing. Nowadays, taxi apps and some entertainment apps can automatically identify the location, the app can be used to send orders to the nearest taxi based on location, or to recommend food and entertainment for customers. Some smart homes can also turn on and adjust the optimum temperature based on the customer's body sensations. This intelligent sensory product not only allows the user's information to be transmitted, but also allows more information to be accessed through sensing, even relying on the customer's five senses, skin and limbs, allowing human organs to be extended.

It is certain that, with the advent of the seventh information revolution, more smart products will emerge; Some products, however, have already been eliminated in a brutal round of fighting before the glory days of the Internet of Things have arrived.

Jawbone was founded in Silicon Valley in 1999 and made a name for itself in the early days as a developer of products and speakers for defence-related projects. In 2011, the company started to commercialise its UP range of health and exercise bracelets. Hosain Rahman, one of the company's founders, is ambitious about the market and has publicly stated that in the future he will create high-tech smart products that are stylish, waterproof, smartphone-connected and have a long battery life. However, its competitors have made bracelets with multiple functions such as screen and sleep status recognition, while the Jawbone bracelet has a single function, each version of the product was uninspired, and with poor hardware and high market prices, it was slow to capture market share. To make matters worse, the bracelet market is also in a slump, with about a third of smart bracelets being abandoned after six months of purchase, according to market research firm Endeavour Partners, making them a veritable chicken-and-egg product. The worst part is that Jawbone's management is in serious trouble: Senior personnel changes prevented the smooth development of product features, and in May 2016, the company announced that it was discontinuing the production of bracelets and moving into the medical field.

The competition for bracelets is actually pretty brutal. When companies like Apple, Samsung and Fitbit entered the field of bracelet development, Jawbone was caught in a brutal competition, while the Chinese brand Xiaomi's bracelet manufacturer also actively joined the development of smart bracelets and led the field alongside Fitbit, accelerating Jawbone's defeat. In 2017, Jawbone announced the commencement of liquidation proceedings. With its energy and money woes and a slew of lawsuits, Jawbone's future is uncertain if it will be able to turn around when the Internet of Things explodes.

The principle of the smart bracelet is not complicated, the core is to obtain health information by sensing the workings of the user's relevant organs. Recording and storing this information provides the user with visual health data. Humans rely on their senses for their first knowledge of the world. The emergence of smart products has helped to extend the human senses even further, and human understanding of the world has become more in-depth and extensive through these sensors. These sensors are constantly sensing and acquiring large amounts of data, which they save and analyse, creating an uninterrupted record and re-analysis, and ultimately Big Data.

Smart sensor devices are not just bracelets, they are just one representative of a wide range of products. Intelligent sensing is about taking the perception of the world from the human organism to the machine, and extending it further and stronger. The five human senses—eyes, ears, nose, tongue and mouth—can be simulated through intelligent sensing. We can't see formaldehyde, we can't smell TVOC (one of the more serious effects of the three types of pollution affecting indoor air quality), but we can sense it through intelligent sensing. In addition to everyday life, the quality of the atmosphere, the quality of water, whether a mountain is about to slide or whether a manhole cover has been displaced ... can also be sensed by intelligent sensors. Through these sensors, human perception of the world will transcend the limits of distance, size and even our five senses, and they will become an important guarantee of human intelligence.

Big Data Rebuilds the Foundations of the Cognitive World

First of all, it is important to note a misconception: data is not a number. The scope of data is much larger than numbers. Wu Jun, former Vice President of Tencent, described Big Data in his book "The Age of Intelligence" as follows: any content on the Internet, such as text, images and videos, is data; all files in the hospital, including medical images, are also data; the various design drawings in the company and in the factory are also data; The texts and diagrams on the excavated objects, even their dimensions and materials, are also data; even the formation of the universe has left behind many data, such as the number of elementary particles in the universe.

It is easy to see that in the 5G era, with high speed and low latency information transmission, human use of big data will help deepen our knowledge of the world, and on the basis of this knowledge, organisations and individuals at the forefront of the times will develop more applications to meet people's various needs and services. With the arrival of 5G, a large number of IoT applications are being used, these IoT sensor devices generate huge amounts of data every day, These figures go far beyond the statistics and management that we do today on a daily basis.

In fact, big data has been used by national authorities since before the Internet became commercially available. One of the earliest and most typical applications is CCTV's weather forecasts, which are based on the recording, modelling and analysis of big data. As massive amounts of data are utilised in the 4G era, the general public can already feel the benefits of big data.

As Internet technology continues to evolve, the use of big data is now even faster and easier. Weather information that we previously needed to see on TV can now be seen on our mobile phones. Unlike the weather forecast on TV, the information we see on our mobile phone screens is much richer. In addition to the temperature on a particular day, we can also see the temperature and weather conditions at various times of the day, or even for longer periods such as seven days or half a month, as well as travel and clothing guidelines. In addition, GPS positioning used to be used only by government departments, but nowadays it has already been put into commercial use, and various mobile phone software and mobile phone servers have positioning functions. All of this is the result of the mature application of big data.

In 1997, at the 8th IEEE Conference, Michael Cox and David Ellsworth presented their paper "Application Control Program Request Page Scheduling for External Memory Model Visualization", in which they first used the concept of "Big Data". In 2000, the University of California at Berkeley published its findings: the quantification of the raw information generated by the four main physical media (paper, film, CD-ROM and disk) each year by computer storage, i.e. all information can be expressed in units of computer memory. With the volume of data expanding at a rocketing rate every year, "big data" has officially become synonymous with massive information flows.

The concept of big data is now in full swing, in large part because of the rapid development of the smart Internet. The powerful Internet provides it with nourishing soil and a runway to soar. Big Data in the 5G era will rebuild a new social order and the basis of human cognition.

The onslaught of big data has not only fuelled the information revolution, but has also facilitated the two-way communication of information. In the case of the most commonly used Internet taxi service, taxi drivers were not friendly before the advent of taxi software. They simply carry passengers and it is a one-off deal, not a long-term model. Thus, once the driver has delivered the passenger to the destination, the transaction is complete and there is no additional service: if the customer is carrying luggage, most drivers do not provide the appropriate service if the customer is elderly, and passengers do not usually complain about such trivial matters, which means that: As there is no evaluation mechanism in place to control the drivers, and no data is accumulated at a later stage, the performance of the drivers is not bound.

However, in the age of big data, all the details of taxi rides and passengers will be stored as data. In this network of people and things, ride-hailing software provides a portal for all drivers and passengers to evaluate: both the passenger and the driver can evaluate each other at the end of the transaction, and anyone else can see these historical evaluations. How a passenger evaluates a driver will have a direct impact on that driver's performance. Ride-hailing platforms will reward or punish drivers according to their rating (in May 2018, the rating function of an online ride-hailing software was shut down after an incident in which an air hostess was killed by an online ride-hailing driver). In this environment, the drivers who join the taxi system make a shift in their service. At the same time, the driver can also see a customer's credit history and choose whether or not to take an order. Technically speaking, all ratings and reviews on taxi platforms are big data with high commercial value.

The book "The Age of Intelligence" states: big data's three main characteristics are massiveness, multidimensionality and comprehensiveness (or completeness). We can get a good sense of these characteristics by searching for keywords through search engines. When these characteristics are used to their full potential in different areas, they are bound to have unexpected effects. Ideas lead the way, and the British Industrial Revolution happened for a reason, It is because people's way of thinking has changed that all kinds of mass-produced machines have been created. In the current era of big data, the flow of information is a major factor in the flow of data, People's way of thinking has been shaken to the point of reversing their perception of the world, this impact and disruption is also fast pushing us to re-perceive the world.

Big data has brought about a huge change in our understanding of traditional shopping, the most profound experience is e-commerce, and the powerful logistics technology behind it has overturned people's perception of traditional shopping.

When online shopping entered people's daily lives and became widely accepted, a variety of shopping websites sprang up. Unlike many shopping platforms, a key to Jingdong's success stems from its self-developed and established set of logistics technology, which encompasses the entire shopping and delivery process and the entire value chain: from the front end of the transaction, to the product supply chain, and then into the core storage, distribution, customer service and after-sales system, and ultimately down to each user's shopping and browsing records. The data accumulated in this series of processes is the basis for the application of big data in Jingdong. In 2008, Jingdong's technology team was just over 30 people, but in just a few years it has grown to over 4000, including a big data team of over 300 people. What's even more surprising is that Jingdong's technical team can even determine a customer's shopping mood based on the storage of big data, so that the customer is equipped with customer service that is good at dealing with the relevant emotions. Today, the large database has become the heart of Jingdong, non-core technical staff can not get close, want to enter Jingdong internship are required to pass a number of hurdles. The use of large databases by the technical team has given Jingdong's logistics a crushing advantage over other logistics companies and has widened the gap with other logistics companies.

At present, the use of big data by some shopping sites is still at an early stage, dominated by one-way output, dividing products according to themes: the range of products and advertising categories are promoted in a single line; and Jingdong takes a multi-pronged, customer-led approach: in addition to the thematic divisions and promotions of the products, users can see them in Jingdong, Jingdong's technology team also creates a data bank for each customer that belongs only to that customer. For example, browsing history, items ordered, when to cancel an order, whether to buy again, etc, all the details of each customer's actions on the Jingdong website are retained and stored in full, forming a complete data chain.

The second aspect of the use of Big Data in Jingdong is to divide all customers into different groups and to push different coupons, related services and products to different groups. The third aspect is to make predictions based on the data left by the user, even to the point of predicting exactly which menu the user will click on next. In addition to the analysis of the user side, the application of big data also directly determines the operational aspects of the company: whether the opening of distribution sites and pick-up points can fully cover a certain neighbourhood and whether the cost efficiency of the entire logistics chain can be improved, etc. A strong accumulation of data also has a direct impact on the decisions made by the top management of the company.

It is worth mentioning that Jingdong has developed a different model of cooperation with its product partners. Traditional suppliers usually do market research first, then develop the product, test it, mass produce it and finally distribute it to retail shops for promotion. Jingdong will open up the relevant database to its partners, so that they can determine the purchasing needs of customers based on the database and use it as a basis to decide the direction of the next generation of product development to meet the expectations of most customers. By working together in this way, product suppliers are able to significantly reduce costs and increase the efficiency of research and development. Using data to speak has gradually become the most important operational strategy of Jingdong. The successful application of Big Data in the field of e-commerce is a testament to its great carrying capacity and the successful transformation of people's mindset.

The high carrying capacity of big data is of great importance to social development and to people's work, study and life. In particular, the massive, multidimensional and comprehensive nature of Big Data is destined to give rise to something new, and that something is intelligent learning above all else.

Intelligent Learning Makes It Possible for Machines to Outperform Humans

If artificial intelligence is the ultimate product of intelligent learning development, then intelligent learning based on big data is the basic capability of artificial intelligence. Memory is the most fundamental function of intelligent learning. Anyone who has been driving for a long time will have experienced this: if you travel frequently to a destination, there is usually a more convenient and closer route based on your own experience over the years. At first, such routes are not shown on the smart navigation, which usually gives routes on the main roads known to the public. If the driver does not follow the navigation route every time, but takes a more familiar path, the intelligent navigator will, through intelligent learning, make intelligent corrections and optimise the previous route by remembering and modifying it after finding out that the driver has chosen such a route every time.

Intelligent learning is based on a large amount of analysis of big data, and through the summary of the optimal choice, to find an efficient, low-cost, convenient and fast path.

The emergence of robots and intelligent products as the endpoints of intelligent learning will also disrupt human society in the future. Humans have created language, invented paper, discovered electromagnetic waves, and performed countless miracles. This time, humans have created robots, but they have to compete with them.

At the end of the twentieth century, the American company IBM built a super computer, Deep Blue, with 32 microprocessors and 480 special chips, using the C language. To achieve such an exciting battle, the scientists entered more than two million games from the world's best players into Deep Blue and then invited chess grandmaster Garry Kasparov to play against Deep Blue in a summit match. From 10 to 17 February 1996, the Man-Machine War began in Philadelphia, USA, and after several days of intense combat, Kasparov defeated Deep Blue by 4:2. A year later, an improved Deep Blue was back in action, now twice as fast as the previous year, able to search and estimate 12 moves after each one, and capable of 11.38 billion floating point operations per second. In May 1997, the man-machine battle resumed, with Deep Blue defeating Kasparov by a score of 3.5:2.5, marking a monumental day in the history of the machine world.

However, Deep Blue relies on computation, not intelligent learning, to make optimal choices based on a multitude of inputs.

One of the most exciting machines, destined to go down in history as a match for human intelligence, is the famous "AlphaGo".

In 2016, DeepMind, a subsidiary of Google, used the principle of "deep learning, two brains" to develop the artificial intelligence machine AlphaGo. Unlike Deep Blue, AlphaGo uses a multi-layer artificial neural network and Monte Carlo tree search to learn from its own deep intelligence, based on a move selector and a game evaluator. Unlike previous machines, AlphaGo focuses on Go. In terms of the two different types of chess themselves, Go is far more complex than chess, and AlphaGo has about 30,000 times the computing power of Deep Blue.

In March 2016, AlphaGo took on South Korea's top player Lee Sedol and won easily with a record of 4:1, causing a lot of excitement in the world of Go, and finally Lee said at the Davos Forum that the coldness of the machine made him "feel like I never want to play with it again". The following year, at the Wuzhen Weiqi Summit in China, AlphaGo again went out and defeated the world's top ranked player Ke Jie 9dan with a brilliant 3:0 score, and directly made Ke Jie cry. As a result, the Go community has recognised that AlphaGo's game has surpassed the highest level of human Go masters, and Ke Jie even said that "it is the God of Go and can beat everything".

In October 2017, DeepMind launched an upgraded version of AlphaGo, codenamed AlphaGo Zero, which uses new artificial intelligence technology to turn the zero-based AlphaGo Zero into a top Go master in just three days of training. In December of the same year, at the 4th World Internet Conference, Google CEO Sundar Pichai said that the team was working on the AlphaGo tool, which is a learning tool for anyone who wants to learn Go, and that he himself was using it to learn.

It looks like the lone AlphaGo is about to teach humans how to play Go.

The domination of the chess world by one AlphaGo alone does not seem to be enough to prove the possibility of intelligent robots surpassing humans in the future. Today, all industries are looking into artificial intelligence to find ways to revolutionise their industries. Looking ahead, will humans still be able to control increasingly intelligent robots?

For example, the healthcare sector. In 2012, Google held a science competition and the title was awarded to a high school student who, using a machine with data from 7.6 million samples of breast cancer patients, devised an algorithm to biopsy patients to determine the location of breast cancer cells with an accuracy rate of over 95%, surpassing that of professional doctors.

Whereas previously machines needed to be operated manually by human hands, robots programmed with artificial intelligence cannot only be operated remotely, but can also perform complex activities over long periods of time. In the complex and sophisticated field of medicine, it would be an underestimation to think that intelligent robots can only perform simple consultations and examinations. Artificial intelligence, developed in the explosion of big data, will create one impossibility after another in the 5G era.

One of the impossibilities created is that of surgery in medicine.

At the beginning of the twenty-first century, after a long period of research and development and repeated trials, Intuitive Surgical in the USA proudly launched the Da Vinci (i.e. Da Vinci) Surgical System and in the following years continued to improve the product to gradually master various complex surgical procedures such as gallbladder removal and fundoplication. The endpoint of this surgical system is an artificially intelligent machine with a dual doctor's console, which uses a swap control method to allow the doctors to jointly control the robot's instruments to assist in surgery. Unlike humans, each arm of the surgical robot is much more agile than a human, and the surgical incisions are very small and highly accurate. Through its own intelligent learning mechanism, it can also automatically improve itself as the volume of surgery increases, giving it greater stability than a human surgeon. To date, more than 3000 da Vinci surgical robots have been equipped worldwide and more than 30 million procedures have been successfully performed, with a few hospitals in China introducing them.

Whether it is human-machine combat or human-machine cooperation, the huge opportunities that lie ahead have long been known. Want to revolutionise industry in the 5G era with artificial intelligence, the research and development of artificial intelligence has become a matter of urgency, with the US and Europe already jumping on the bandwagon and China also jumping on the bandwagon.

After AlphaGo's battle against human Go masters, intelligent college entrance exam robots were unveiled in Beijing and Chengdu in the summer of 2017, one is Aidam, developed by Learning to Learn, and the other is AI-Maths, an artificial intelligence system developed by Quasi-Star Cloud Learning Technology Ltd. Both robots took less than 10 minutes and 22 minutes to answer all the exam questions, scoring 134 and 105 respectively, starting with the subject of mathematics. The difference between the two is that Aidam works similarly to Deep Blue in that it deconstructs the language of the test questions, searches through the input knowledge, and then extracts the relevant knowledge from the questions to find the optimal solution; AI-Maths, on the other hand, is similar to AlphaGo, in that it uses integrated logical reasoning and continuous intelligent training to solve problems. Although the development of examination robots is still not mature enough, in other subjects is still a blank, but the future of artificial intelligence can subvert the traditional education industry, let us wait and see.

The autonomous learning approach of artificial intelligence, which greatly surpasses the preferential approach of computers, is set to revolutionise and qualitatively change in the 5G era.

The Smart Internet will integrate the capabilities of mobile connectivity, intelligent sensing, big data and intelligent learning to form a new capability that can penetrate into every corner of social life and influence and change the course of the world.

The Basic Spirit of the Smart Internet

Every industry, every entity and every individual has its own spirit, and the Internet is no exception. The spirit of the traditional Internet is different from that of the Smart Internet, which has evolved from, but is completely different from, the traditional Internet.

The Basic Spirit of the Traditional Internet Is Freedom, Openness and Sharing

Freedom is one of the fundamental spirits of the Internet.

First of all, it is important to emphasise that, from a materialistic dialectic perspective, everything has two sides, positive and negative, good and bad, and can be interchanged. Freedom is not absolute, and like the tails of a coin, freedom and control go hand in hand. With the explosion of the Internet, the transmission of information to and from humans has developed more freely than ever before, compared to the slowness and isolation of the past.

The development of the Internet has enabled human information exchange to break through the limits of time, space and scope of activity. We can search for as much information as we want, and this information is not limited by national borders. On the Internet, Internet users can do whatever they want legally and ethically, whether it is browsing the news, watching videos or writing blogs. In addition to this, people's opinions on the Internet are more diverse than in the past. On many well-known knowledge sites, such as Douban and Zhihu, we can ask all kinds of questions we want to know and answer any questions we want to answer from other users; We can discuss and comment on a topic, a book or a film, delete and edit what we have said, and say whatever we want, regardless of our status. This freedom of network has greatly facilitated the transmission and development of information.

Openness and freedom go hand in hand on the Internet.

On 19 January 2018, Indian national treasure Aamir Khan met Chinese audiences with his new film *Mystery Superstar*, which sparked widespread discussion. In the film, a young girl who has suffered domestic abuse for a long time and is forbidden to sing becomes an Internet sensation by uploading a video of herself singing on YouTube, which in turn changes her fate. This is a typical example of the open spirit of the Internet, which has enabled the emergence of many new industries and contributed to the diversity and sharing spirit of the Internet.

YouTube is the most iconic, giving rise to the new word "Netflix".

The year 2004 was marked by two negative global events, the American Super Bowl indecency and the Indian Ocean tsunami. These two headlines gave former American Online Payments employees Chad Hurley, Shijun Chen and Judd Karim the idea of creating a video site for sharing among friends. In 2005, YouTube was launched on Valentine's

Day, followed by the release of a 19-second video called "I'm at the Zoo". The story that followed is well known.

A billionaire called YouTube a company that "only idiots buy", yet in just one year, YouTube was acquired by Google for $1.65 billion and has since gone on to become a hit, with American singer Justin Bieber, Korean musician Psy, Vietnamese beauty anchor Michelle It has won the Peabody Award, the highest award in the field of electronic media, and has become a media venue for presidential elections, fulfilling its great mission of opening up and sharing information globally.

The openness of the Internet has opened the public's eyes to an era of information inclusion, and its sharing of information has dramatically reduced the sense of space in the world and given people a deeper perception of the world. On this basis, the future smart Internet will face even higher requirements. The basic spirit of the traditional Internet no longer covers the smart Internet, but rather the need to build a comprehensive security system, as well as the management of a strict, efficient and convenient user experience.

The traditional Internet was built at a time when information transmission was not smooth enough, and it was the belief and basic idea of the traditional Internet to break the limitation of information transmission and obtain efficient information transmission through the system of the Internet. At the beginning of the Internet, the phrase "Nobody knows you're a dog on the Internet" was widely recited, and the Internet was tasked with breaking down the old system of information transmission. It is this fundamental spirit that has allowed human society to enter a revolutionary era of information transmission.

The Spirit of the Smart Internet Is Security, Management, Efficiency and Convenience

The traditional Internet is based on a fixed network, with the PC as the main terminal, and in an era of poor information exchange, this network can break down barriers and achieve high-speed information transmission.

The main terminal in the mobile Internet era is the smartphone, location and mobile payment are the most important functions, and the

Internet is gradually moving from information transmission to life services.

In addition to smartphones and PCs, a large number of smart terminals will be added to the Internet, and communication functions will be added to a large number of ordinary devices in social life, so that the Internet will penetrate every corner of social life. Its capabilities are not limited to information transmission, life services, social management, but even penetrate into the organisation of production. The basic spirit of the Smart Internet in this new era is therefore no longer covered by the basic spirit of the traditional Internet.

While the traditional Internet has brought new and unprecedented experiences to people's work, study and play, it has also had a negative impact. The traditional Internet has been a source of controversy in terms of security and regulation, with a lot of negative news such as personal information leakage and account theft. In 2008, the entertainment industry was paralysed by the indecent photo scandal that shocked the nation, as personal information was exposed incessantly due to poor Internet regulation; The outrage of countless people has caused great physical and psychological harm to many artists, and the rapid transmission of this information has undoubtedly violated the sacred law and sounded the alarm for the world.

In May 2017, after "Panda Burning Incense" swept through China, the cyber virus "WannaCry" swept through the world, affecting hundreds of countries and thousands of systems, affecting universities, hospitals, police stations and other important departments of the country. The virus has affected universities, hospitals, police stations and other important departments of the country, crippled medical systems in the UK and infected the Chinese school intranet, with serious consequences.

Unlike the traditional Internet, freedom, openness and sharing will no longer be core values in the age of the smart Internet. Regarding the core values of the Smart Internet, I define them as four spirits: security, management, efficiency and convenience.

1. Security: If the network is not secure, then it is worthless. This is because a large number of applications and services of the Smart Internet are closely related to social security and safety of life. Imagine

if an intelligent transport system is unsafe, an intelligent health management system is unsafe, it may cause more damage than it contributes, therefore, security is the first requirement of the intelligent Internet. The network must be secure for it to be worthwhile.

2. Management: In the traditional Internet era, we may feel that management is of little value and should not be managed by the government, or even managed at all, and that the Internet should be free and open. Is this really the case? Take the Facebook "user information breach" for example. The British data analysis company Cambridge Analytica obtained information on 50 million Facebook users in the run-up to the 2016 US presidential election and, on the basis of this information, managed to help Donald Trump win the US presidential election. At this level, regulation is very important and needs to be managed by a well-developed legal and administrative means. And with a large number of cameras, driverless cars and smart homes joining the smart Internet, management is a fundamental tool to ensure that this system works properly.
3. Efficient: 5G can greatly improve the efficiency of society. Take transport as an example. In some places, serious traffic congestion is now caused by the high volume of traffic at certain times of the day. Intelligent transport will create an efficient transport system, the data is collected and then allocated appropriately. The most advanced form of intelligent transport is one in which all vehicles are controlled by a central control centre, e.g. at what speed and on which road.
4. Convenience: Compared to the traditional Internet, the Smart Internet will make our lives easier. 4G has already brought m-commerce, bike-sharing, car-sharing, takeaway services and mobile payments into people's lives. 5G-based smart Internet will greatly improve the efficiency of all social life, making users feel more convenient and faster than ever.

Specifically, security is key if the new information revolution is to gain full acceptance by the public. From the current development trend, the future will be the world of mobile Internet, and ensuring the security of personal information is the first problem to be solved by the smart Internet. While the main task of the traditional Internet is to fulfil the mission of information transmission, the Smart Internet needs to be better and more deeply involved in the entire social life system, including intelligent transportation systems and health management. Imagine if the intelligent traffic data was hacked, the whole city would be in a real-life version of a driverless "life and death situation", which would undoubtedly be a disaster. If the personal health information of a major figure is compromised, it is not just the individual but the entire interest group that suffers. While there is a great deal of expectation for a smart Internet, there are also greater demands on security, which is why a more sophisticated network management system is needed.

Just as openness and freedom co-existed in the traditional Internet, security and management will inevitably coexist in the future era of the smart Internet. There are many loopholes in the traditional regulation of the Internet and the regulatory system is not perfect. In the 5G era of mobile connectivity, there will be an unprecedented explosion in the transmission of information and ubiquitous networks. If we want to ride on this revolution, a strict and comprehensive management system must be established, known to the public, to convey the spirit of information management and security, otherwise, it will easily lead to serious social problems.

In addition to intelligent transport, the future of intelligent healthcare will have a large amount of personal information and treatment progress data of patients, how to carry out treatment, how to track the condition over time, if the information related to these issues is not managed, then this private information is likely to be leaked and used by unscrupulous people for other purposes, many operations of medical institutions will not be able to carry out, the whole healthcare system may face paralysis.

If security and management are the basic spirit of the Smart Internet, then efficiency and convenience are its core spirit.

The mobile Internet and big data will explode in the 5G era. With a complete security and management system in place, the efficient operation of the smart Internet will be intuitive to all users, and the gears of the entire society will move forward as if a high-speed motor has been installed, running non-stop to achieve the missile-like transmission of information. At the same time, the increase in social efficiency will make life easier for people. Once artificial intelligence is in full bloom, smart living doesn't need to be managed manually at all, with sophisticated sensors and smart homes that can be switched on and adjusted without wasting any extra time in operation.

2

What Is True 5G?

The Evolution of Mobile Communications

When viewed over a longer time scale, human communication began as face-to-face communication, the earliest form of long-distance communication was the use of wolf smoke to convey news of an approaching enemy, while a more sophisticated form of communication was the writing of words on paper, which was transported to others by post. Post communication has been the most advanced communication system for a long time, but it has obvious drawbacks: it does not allow for real-time communication.

The first real time communications to emerge were the transmission of basic information by telegraph and later by fixed line telephony. Compared to telegraphic communication, which requires a coding and decoding process and has a lag time, telephony can deliver information directly, but whether using copper or fibre optic cables, fixed line telephony is limited by the scenarios in which it can be used, so we need to find a new technology that allows real-time communication anywhere, anytime.

L. Xiang, *The 5G Era*, https://doi.org/10.1007/978-981-99-4534-4_2

1G: Humanity Enters the Age of Mobile Communications

The idea for the first generation of mobile communications was presented at the 1939 Universal Exhibition by AT&T, the largest telecommunications operator in the United States at the time. The idea was rejected by the Federal Communications Commission (FCC) because, just as you need land to build a house, you need spectrum for mobile technology, and at the time the right spectrum was in the hands of scientific institutions, the military, the police, radio and television stations, and so on.

The matter was put on hold for 30 years. In 1969, television technology made the transition from wireless to cable. Because wired transmissions were more stable and of higher quality, many stations returned their spectrum, and it was then that the spectrum became available for mobile communications. The Federal Communications Commission, the US communications regulator, thought that the spectrum could be used for mobile communications, which had been on hold for decades, and this is what started the push for AT&T to develop mobile products.

At this time, another company in the United States also began research into civil mobile communications, this company was Motorola, which had some technological accumulation in the field of military communications at first. Martin Cooper, head of Motorola Mobility, asked the technical department to design a mobile phone in 45 days, and this was one of the world's first concept phones.

It was not until 1973 that the mobile phone was officially established. At that time, a base station could only support four people at the same time, and the phone could only make a call when it was connected to the base station.

At this point, new technologies and approaches need to be found to support more people on the phone at the same time. A very important event occurred when AT&T successfully established a cellular mobile network and launched the first commercial trial of an analogue cellular network in Chicago using AMPS technology. What is cellular technology? A single base station covers a hexagonal cellular area, and each base station uses the same spectrum, using spectrum multiplexing, so that the spectrum can be reused, so that a small amount of spectrum can be reused by everyone.

It was at this point that the telephone became truly commercially viable. The world's first commercial mobile network was established in Japan in 1979, followed two years later by the establishment of cellular mobile networks in Bahrain and Northern Europe. Because mobile communications do not require a lot of wires to be pulled to each home and then connected using a switchboard, the cost of building mobile communications is much lower than the cost of a fixed line for emerging economies.

Interestingly, however, mobile communications had not yet begun to be built in the United States. The Federal Communications Commission has spent years reviewing this in the hope of finding a fully level playing field for mobile communications, based on the concept of market competition. In March 1981, AT&T and Motorola were still waiting in the wings.

In terms of time, the Americans were the first to research cellular mobile systems, but it was other countries that were the first to use them, which the Americans found unacceptable and even "humiliating". However, the real solution to this problem requires some skill. What to do? In the end, they also resorted to networking and "backdooring" people.

In hindsight, the way it was handled was very interesting. Bob Galvin, the CEO of Motorola, was an old friend of the then Vice President of the United States, George W. Bush, and once called him to say he was taking his young grandson to Bush's office. While chatting in the office, Calvin asked Bush: "have you ever seen a portable phone?" Bush said: "No, I haven't seen that."

Calvin then took out his mobile phone and gave it to Bush, he asked Bush to call his wife, Barbara, and Bush spent some time talking to her on his mobile phone. It was clear that Bush was impressed by the new product and, for the first time as a consumer, he felt that a mobile phone could be so convenient that he suggested that Ronald should see it too. Calvin didn't respond yet and asked which Ronald, and Bush said it was Ronald Reagan, the president at the time.

Calvin then went to see President Reagan, who looked at this phone and asked: "How's this thing coming along?"

Calvin said, "We could have gone public a long time ago, but His Excellency's Federal Communications Commission wouldn't let it move."

Reagan turned to his secretary and said, "Give them a call and tell them to issue the listing permit immediately."

Two months later, the licence, which had been delayed and discussed for eight years, was finally issued and the United States was able to enter the mobile era. At this time, the United States was two years behind the world's first commercial cellular mobile networks.

In this way, mankind began the era of mobile communication. At that time we did not have the concept of generation, but nowadays it is the first generation of mobile communication based on analogue communication, which is known today as 1G.

Cellular mobile communications were undoubtedly a major revolution in the history of mobile communications, with analogue cellular systems being widely used in Europe and the Americas throughout the 1980s (see Table 2.1). The pace of mobile communications in China has been a little slower than in the West and developed Asian countries. On 18 November 1987, China's first TACS analogue cellular mobile phone system was

Table 2.1 Main cellular systems present in Europe in 1991

Country	Systems	Frequency band	Establishment time	Number of users / million
UK	TACS	900	1985	120
Sweden, Norway Finland, Denmark	NMT	450,900	1981–1986	130
France	Radiocom2000NMT	450,900	1985–1989	3 009
Italy	RTMSTACS	450,900	1985–1990	6 056
Germany	C-450	450	1985	60
Switzerland	NMT	900	1987	180
Netherlands	NMT	450,900	1985–1989	13
Austria	NMTTACS	450,900	1984–1990	606
Spain	NMTTACS	450,900	1982–1990	606

completed in Guangdong Province, and at the same time, China's first mobile phone bureau was opened in Guangzhou, the first generation of analogue mobile phones entered China, the expensive and bulky mobile phone was called "Big Brother".

In the 1G era, mobile phone voice transmission was realised and our mobile phone public network consisted of the Motorola mobile communication system (A system) in the USA and the Ericsson mobile communication system (B system) in Sweden, i.e. the A and B networks. There was no connection between the A and B networks, so mobile phones were not allowed to roam at that time. If you are in Beijing and going to Shijiazhuang on business, you will be connected because both places are on the A network; if you are in Beijing and going to Chengdu on business, you will not be able to get through because Chengdu is on the B network. The mobile phone is also known as a "brick phone" because of its heavy and bulky body, which is limited to voice calls.

At that time, the mobile phone market was dominated by Motorola and Ericsson, most notably the Motorola 3200, which appeared in Hong Kong police films at the end of the twentieth century, and the 8000X, which was on the market. After the introduction of this batch of Big Brother models, the supply in Guangzhou immediately exceeded the demand, and the price of a mobile phone was more than 20,000 yuan, which was a rarity for the ordinary people with an average monthly salary of less than 100 yuan, as a result, the mobile phone was largely a symbol of wealth and status in those days, and the first mobile phone users were limited to business people and senior government officials.

In 1987, the Guangzhou Post and Telecommunications Wireless Branch took the opportunity of the National Games to launch the first 100 analogue mobile phones on a "pilot" basis. A young man in his 20s, Xu Feng, became China's first mobile phone user with the help of a relative who worked at a post office. According to the person's recollection, the person at the post office did not know the price of the phone and only asked him to deposit a cheque for $20,000. After depositing the cheque, Xu Feng was given a NEC analogue mobile phone and had to pull out the antenna to make a call. With excitement and curious eyes, Xu Feng made his first call to Hong Kong on his mobile phone, to the surprise of everyone around him. Although there was no so-called roaming in the first six months, it was enough to make Xu Feng proud. In the 1G era of analogue mobile communication, the quality of calls on mobile phones was not good, and it was often necessary to speak loudly into the phone, which drew the attention of people around us, followed by admiring glances. To buy such a phone, you have to go through all sorts of relationships with operators, stand in long queues, and repeatedly file application reports and stamps. In 2016, Xu Feng, already chairman of Guangdong Zhonghai Group, was interviewed by a reporter, when I think back to the first time I got a mobile phone, my face is still full of pride.

The technology used in the first generation of mobile communications was mainly analogue. By analogue communication, we mean turning our sound into electric waves, transmitting it through the waves, and then reducing the waves to sound. So the first generation of mobile communications has poor quality, poor security, susceptibility to interference and poor spectrum utilisation, but it has established the most basic capabilities of mobile communications, such as cellular communications, spectrum multiplexing and other core technology tools.

The 1G era solved the basic problem of mobility, but in the years to come, the mobile industry will have to undergo a series of changes. While analogue cellular technology and brick phones have surprised people, they have also revealed serious drawbacks: Analogue technology suffers from low capacity, mobile phone theft is rampant and expectations of price, call quality and off-site or cross-border roaming come with the realisation of mobile communications … however these cannot be met by 1G. In 1999, the A and B networks were officially closed and digital communications came into being.

2G: The Arrival of the Digital Age

The key differences between the technologies of the second generation of mobile communications, which have gone further than the first generation of mobile communications, are, it first turns the sound information into a digital code, transmits it through the digital code, and then uses the other party's modem to decode the code and demodulate it into sound, so the second generation of mobile communication is stable, anti-jamming and secure. Because of the digital encoding technology, it is possible to achieve things that were not possible in the 1G era, such as: caller ID, call tracking, SMS, etc.

More importantly, since the second generation of mobile communications, there has been a worldwide competition for mobile standards, in which the International Telecommunication Union (ITU) has played a very important role.

On the surface, the communications industry is a collective struggle between major telecommunications operators and mobile phone vendors, but in reality it is a fierce competition between countries for strategic soft power, once you have seized the opportunity, you will be able to make a good turnaround and take the initiative for a long time, dominating the industry. While the first generation of mobile communications was monopolised by Motorola and dominated by the Americans, the second generation of mobile communications has seen a hundred rivalries, with several powerful economies around the world developing their own standards.

The inventor of CDMA (Code Division Multiple Access) technology, Hedy Lamarr, must be mentioned when talking about second-generation mobile technology in the United States.

Hedy Lamarr was a communications student who was so beautiful and glamorous that she gave up her communications degree to become a film star. She began her acting career at the age of 16 and two years later, at the age of 18, Heidi was cast in the lead role in Ecstasy.

Hedy Lamarr came from a privileged family, her father was a banker and her mother a pianist, but she didn't want to live like a traditional lady and chose a different path. At the age of 20, Rama married Fritz, then a prominent Austrian arms dealer who manufactured arms for the Nazis, particularly flight control products. The intelligent Rama learns from her husband about cutting-edge ideas in communications technology, including military classified communications. In 1937, Hedy Lamarr, who was

Jewish, decided to leave her husband after the German-Austrian merger. At a dinner party, she drugged the maid while her husband was busy socialising and escaped by train to Paris, he then moved to the United States, where he was discovered by an MGM director and officially entered Hollywood.

In her haste to leave, Hedy Lamarr fled her husband with nothing, but she had a "priceless treasure" in her head that she brought to the Allies as a "military secret" in Nazi wireless communications. These secrets are mainly based on "command-guided" systems with secure radio communications, which are used to automatically control weapons and strike targets with precision, and to prevent radio commands from being stolen by enemy forces.

In the early 1940s, Heidi met the musician George Ansell in Hollywood. George Ansell also hated the Nazis and Heidi approached George with the idea of creating a secret communications system, wanting to develop a military communications system that would stop enemy radio waves from jamming or eavesdropping. Drawing on the piano with which Georges was familiar, a model of a device capable of automatically compiling codes was developed, following Heidi's idea. Through the ingenuity of the two men and the help of other scientists, they completed the study. On 10 June 1941, the pair applied for a patent for the frequency hopping technology for which Hedy Lamarr is also known as the "mother of frequency hopping".

During the Cold War, frequency hopping was used extensively for covert communication products because of the special needs of the space-time environment. After the end of the Cold War, frequency hopping technology was finally declassified and allowed to enter the civilian sector, and frequency synchronisation methods shifted from mechanical to electronic, leading to significant developments in radio communications. In 1985, an American company founded in San Diego quietly developed a CDMA wireless digital communications system based on the frequency hopping technology that is now known worldwide as Qualcomm.

So what do you mean by code division multiple access? Give an example. When we all speak together, some of us speak Chinese, some Japanese, some German, and the only information that people who speak Chinese really receive is Chinese. This is also the case with Code Division Multiple Access (CDMA) technology, where data is packetised into packets and divided into different addresses using different codes, when you receive it, you will only receive my code, it is similar to sending a courier, number one is yours and number two is mine, a standard that is one of the best in the world in terms of quality of communication.

Japan was also doing mobile communications at this time, as the strongest economies in the 1980s were the US, Japan and Europe, so Japan made the PHS standard, which was the source of the technology for SmarTone.

When the wind of digital communication first started to blow, European countries realised the problem and adopted a strategy of huddling together: Learning from the lessons of the 1G era, it is impossible to compete with the United States if we establish standards behind closed doors. For decades, the European Union and the United States have been competing against each other. As early as 1982, the European Postal and Telecommunications Commission set up the Group of Experts on Mobile (French: Groupe Spécial Mobile, which stands for GSM and later changed its full name to Global System for Mobile Communications). The Group was set up in 1982 by the European Postal and Telecommunications Commission to study standards for mobile communications.

In 1991, Ericsson and Nokia built the first Global System for Mobile Communications (GSM) network in Europe and launched it commercially in Finland, marking the official arrival of the second generation of mobile communications technology, the 2G era. A year later, the European Committee for Standardisation introduced harmonised standards, adopting digital communications technology and a unified network standard, and developing more new services for users. The core of GSM technology is Time Division Multiple Access (TDMA), which is characterised by dividing a channel between eight callers, with only one person speaking at a time and each person taking turns using 1/8 of the channel time. This system is easy to deploy, supports international roaming and has a SIM card. The biggest breakthrough in this generation of mobile technology is the ability to send 160-character text messages thanks to the use of digital coding instead of the original analogue signal.

Thus, during the 1980s and 1990s, there were three major standards around the world, each of which developed in its own way. European countries are united in their efforts to seize the opportunity. CDMA in the US started later than GSM and lost half of its territory as soon as it was introduced. At the same time, Qualcomm had no handset manufacturing experience, European operators did not care about its intellectual property, the media did not stand by CDMA, and only a very small number of US operators used the system, so the base stations were not built as expected. In the 2G era, the US CDMA standard lost its dominant position in the 1G era.

The loss of CDMA dominance also had an indirect negative impact on Motorola. In the 2G era, digital mobile phones are gradually replacing analogue mobile phones, and although Motorola's analogue mobile phones still have a 40% share of the market, their share of the digital mobile phone market is minimal. Although Motorola has since launched classics such as the StarTAC, it has not been able to reverse its decline. It probably never dreamed that its monopoly on the 1G era would finally be crushed by a company from Finland that started out by making paper from wood and only launched its first digital phone in 1992—Nokia.

In 2002, Spielberg's masterpiece "Minority Report" was released, the Nokia 7650 was made famous by the film. The phone was innovative, high-tech and had a camera, a slider and a five-dimensional joystick, and was Nokia's first colour screen phone. This bold attempt made the Nokia 7650 famous, with a market share of over 40% and impressive sales. Although Sony Ericsson also had a good run in the 2G era and launched the classic T618 in 2003, the whole mobile phone market was still dominated by Nokia.

Technological innovations have also led to a significant reduction in the cost of mobile phones, which, although still expensive, are no longer a luxury item and are now being brought into the homes of millions of people.

When it came to the introduction of mobile communication standards in China, the then Minister of Posts and Telecommunications, Wu Jizhuan, first rejected PHS because the technology, if used on a large scale, would be more costly, less effective in transmission and less adaptable.

In 1994, China Unicom was established as the first step in the telecommunications reform. A few months after its establishment, Unicom announced the deployment of GSM in 30 provincial capitals in China. Following the announcement, the leadership of the Ministry of Posts and Telecommunications held an emergency meeting in Langfang, Hebei, after which the Ministry's Mobile Bureau announced the deployment of GSM in 50 cities in China, and China's mobile communications officially entered the 2G era, gradually establishing the two largest GSM networks in the world.

Later, during the negotiations for China's accession to the WTO, China Unicom set up a CDMA network as a bargaining chip with the US.

In the 2G era, communication technology has evolved from analogue to digital. Not only is the voice transmitted digitally encoded, ensuring high quality and anti-jamming capabilities, but digital communication capabilities (such as SMS) have also been increased, and digital communication services such as caller ID can be provided, with network speeds reaching 9.6Kbps, and even higher with GPRS technology. With the rapid growth in the number of users, the capacity and speed of 2G has hit a bottleneck, and with the rise of multimedia, 2G technology is no longer able to meet the needs of mobile multimedia development.

A review of the entire history of the 2G era reveals the following issues:

Only the economically and technologically strongest economies can set communications standards; The US, for example, made the adoption of CDMA a key condition of China's WTO accession, while Europe has helped European companies to develop by harmonising standards through international organisations.

The importance of establishing communication technology standards for the economic and technological development of a country cannot be overstated. In the Internet era, all computer standards—operating systems, CPUs, etc. and even the production of computers—were dominated by the United States; in the mobile era, Europe began to catch up, with Ericsson, Nokia, Philips, Siemens, Alcatel and Sagem all developing into very large companies. So, from this point of view, all powerful countries in the world have to do something about communication technology standards.

3G: Coming of Age of Data

From 1G to 2G, communication technology has undergone significant changes and the industry has undergone a major reshuffle in just a few decades. As the old Chinese saying goes, "The wind and water turn," time tells not only a story, but also a myriad of eggs inside. While Europe led the way with GSM and Nokia dominating the 2G era, the already dull American Qualcomm teamed up with the Koreans and quietly rose to the top.

In 1985, Irwin Jacobs, PhD in electrical engineering at MIT, and Andrew Viterbi, the originator of the Viterbi algorithm, sold their electronic communications company in San Diego, California, to form Qualcomm, which commercialised the CDMA technology used in military communications during the Cold War and significantly improved the technology's power problems. Unfortunately, in the 2G era, GSM's TDMA technology was recognised by the US Communications Industry Association as a step ahead of Europe, and although CDMA had high capacity and high-quality calls, the technology was complex and therefore not trusted by operators.

In November 1990, Qualcomm signed a technology transfer agreement with the Electronic Telecommunications Research Institute (ETRI) of Korea for CDMA. CDMA is the only standard for 2G mobile communications in Korea, and Qualcomm pays 20% of the royalties it collects in Korea to ETRI each year. Prior to this, Korea's communications industry was generally very weak, and after the agreement was signed, major brands such as Samsung and LG in Korea received strong support to focus on the commercial use of CDMA.

After five years of development, Korea's mobile subscribers surpassed one million, SK Telecom became the world's largest CDMA operator, Samsung Electronics became the world's first CDMA handset exporter, and Qualcomm became a global multinational with its partnership with Korea's telecommunications industry and turned around perfectly in the 3G era.

The 3GPP (3rd Generation Partnership Project) was formed by European manufacturers in conjunction with Japan and other countries using the GSM standard to develop the third-generation communication standard, WCDMA. Seeing this, Qualcomm rushed to join forces with the Koreans to form the 3GPP2 (3rd Generation Partnership Project 2) to develop CDMA2000.

At the same time, China has taken the first steps towards experimentation. In January 1998, a conference on candidate technology submissions and China's strategy for identifying 3G candidate technologies was held in Xiangshan. At the conference, professors from universities and researchers from research institutes presented some of their views on 3G technology research, and the conference was attended by 20 to 30 people, with 90% of them being sceptical.

In fact, there is a particular background and justification for the scepticism of the experts. Previously, international standards have always been the world of foreigners, when it comes to mobile communication standards, not only the cost is very high, but also difficult, whether China can afford to play this game is an unknown. To put it bluntly, it's like an intense competition where people who have never played before are very apprehensive about entering and trying it out.

In the face of controversy, Song Zhongyuan, then Director of the Science and Technology Committee of the Ministry of Posts and Telecommunications, made the decision: "China cannot always rely on foreign technology to develop its mobile communications business. The first time may not be successful, but it will leave a valuable experience. I support taking TD-SCDMA to the international level. If it does fail, we will see it as a victory, an attempt by Chinese people to be innovative and make a contribution to the country." At this point, the TD-SCDMA conference was the final word on TD-SCDMA.

TD-SCDMA technology was proposed by the Institute of Telecommunications Science and Technology of the Ministry of Posts and Telecommunications—later Datang Telecom. When it comes to TD-SCDMA technology, there is one person who must be mentioned, and that is Li Shihe, the "father of 3G" in China.

While working abroad, Li Shihe developed SCDMA technology—one of the world's first technologies to use smart antenna technology—and achieved some success. After returning to China, he went to Xinwei

Communication Technology Co Ltd, where he focused on the development of SCDMA technology and its application to rural mobile communications. In the process, it was suggested that: "We in China can also develop a mobile communication standard", coincided with the ITU's call for third-generation mobile communication standards, and someone else suggested: "Are we allowed to participate too?"

Li Shihe thought it was a good idea, so he actively discussed it with a number of people, hoping to get some inspiration from it. In the course of the exchange, he met a "valuable person"—Li Wanlin, the head of mobile communication standards at Siemens in Germany. After some conversation, Li Wanlin felt that Li Shihe was a man of ideas and ability and invited Li Shihe to Germany for an exchange.

Because the third-generation mobile communication standard is open to the world, countries are working feverishly on it, everyone has their own ideas and technologies, and Siemens was no exception, when it came up with the idea of using TD as a mobile communication standard.

There are red and green wires in the phone, one wire is responsible for sending the information in and one is responsible for sending it out. Communication is about working in both directions, so how does it work in both directions? Mobile proposes two concepts.

The first is frequency division duplexing, which means that the uplink and downlink transmissions are carried out on different frequencies, which in layman's terms means dividing two different frequencies to work in both directions. For example: The 32.4Hz to 32.6Hz band is dedicated to your signal, and the 32.8Hz to 33.0Hz band is dedicated to my signal, using these two different frequencies to split into two channels, one for outgoing and one for incoming information. The second is Time Division Duplex (TDD), which means that the same frequency is used for incoming and outgoing signals with time as the division point. For example; in the frequency range of 32.4 Hz to 32.6 Hz, messages are transmitted to me at one time and to you at another time, alternating in this way. The advantages of frequency division duplexing are: there are two lines, each responsible for its own part, just like a highway divided into two roads, two directions, very efficient, but the disadvantage is "occupying a lot of space", the utilisation rate is poor, the frequency needs to be paired, and each pair of frequencies also need to have a section between the interval between each pair of frequencies to prevent mutual interference, so it will take up a lot of frequency resources. Time division duplexing, on the other hand, builds a "road", so it takes up less "space" and is therefore less efficient than frequency division duplexing, but if the speed is very fast, for example, if you are travelling back and forth at super speed, then even if you are running on a "road", it does not matter much. In short, the two techniques have their own advantages and disadvantages.

Li Shihe travels to Siemens in Germany for an appointment. This is because he developed the SCDMA technology, which is also very innovative in terms of smart antennas, so Li Shihe proposed to Siemens to combine his SCDMA technology with Siemens' TD to make a new standard. However, at that time the whole of Europe had already decided to adopt WCDMA technology and Siemens did not agree to make a separate standard for itself in the interest of Europe as a whole.

But Siemens said that if the Chinese want to develop a standard on their own, they can do it themselves, and Siemens will provide part of the TD technology, such as development tools, development ideas and so on, as a reference. In this way, Li Shihe, who has received technical support from Siemens, brings his students along to participate in research and development discussions. The TD-SCDMA technology was also gradually shaped by Li Shihe's team's research and development time and again.

30 June 1998 was the last day for the International Telecommunication Union to call for standards, after which date all submissions were invalid. It was on this day that China submitted the standard it had developed. There are two reasons for choosing this date for submission: The first is that the TD-SCDMA standard needs time to be revised and improved repeatedly; the second is that we do not want to submit it too early, so that others know that China has also made a standard. At that time, some self-developed technologies were expected to be kept completely secret from the public.

When he heard that China had also submitted a standard to the International Telecommunication Union (ITU) in Geneva, Switzerland, the Chinese Director of the ITU's Bureau of Standardisation, Houlin Zhao (now Secretary-General of the ITU), was delighted and immediately went to see what was going on. This looks like a problem, the signature of the standard is actually Beijing Xinwei Communication Technology Co., Ltd. Zhao Houlin immediately contacted the relevant people and told them that they had to submit it in the name of the Ministry of Posts and Telecommunications of the People's Republic of China. Because Xinwei Company has no right to submit the standard to the International Telecommunication Union, if it is submitted in this way, the standard will be invalidated. However, the deadline was already reached and any further submissions after that would have been invalid. Fortunately, there was a seven-hour time difference between China and Switzerland, so the TD-SCDMA standard, signed by the then Chinese Minister of Posts and Telecommunications, was resubmitted within the required time frame.

The main third generation mobile communication standards submitted worldwide at that time were: CDMAEVDO in the USA, WCDMA in Europe and TD-SCDMA in China.

The conference to discuss the third-generation mobile communication standard started in 1998 and ended in 2000, which lasted for nearly two years. The idea was to unify the world's mobile communications to a single standard, but in the process, China and the United States were at odds with each other, and Europe was dominated by the large number of countries and high vote share, while the United States had a low support rate, so naturally the voice was not enough, so in the end it almost became a European "world". The U.S. believes that in the end, the European WCDMA standard will most likely be selected, and of course it cannot accept such an outcome, so the U.S. delegation approached the Chinese delegation and said: "Instead of letting European standards dominate, the US and China should join forces to support each other and turn the situation around." In this way, thanks to the mutual support of the United States and China, the third-generation mobile communication standard has gone from a "near-monopoly" to a "three-legged race".

In May 2000, the International Telecommunication Union (ITU) officially released the third generation of mobile communication standards, with China's TD-SCDMA, Europe's WCDMA and the United States' CDMA2000 together becoming the three mainstream technologies in the 3G era. With the advent of the 3G era, mankind has ushered in the era of the smartphone.

In 1996, Microsoft released the first smartphone operating system, Windows CE, but due to the lack of practical experience on mobile terminals, its system speed was very slow; In 1998, the British company Pison formed Symbian, a joint venture with Nokia, Ericsson and Motorola, to develop a mobile phone operating system to combat Microsoft. However, at the critical period of technological change, the deep-rooted old-fashioned and traditional Europeans held back. For the entire five years before 2004, Nokia still focused on traditional mobile phone functions, was very conservative, and did not listen to any development. Feature suggestions, not to mention touchscreen and app ecosystem development.

While Symbian and Microsoft are fighting it out and Nokia continues to dominate the mobile phone market, Apple is borrowing technology from both phones and has acquired FingerWorks, a company that develops touch technology. In January 2007, Steve Jobs launched the first generation of the iPhone.

The iPhone 1 became famous thanks to a variety of key applications, a simple interface, touch screen technology and the unified platform of the App Store, which also defeated Symbian, which had taken seven years to develop, and became a major turning point in the history of smartphone development, with the smartphone market exploding after 2008.

In May 2008, China Tietong was merged into China Mobile. In June of the same year, China Unicom began its merger with China Netcom, and China Telecom acquired Unicom's CDMA network for a total of RMB 110 billion. On 7 January 2009, China finally issued three 3G licenses, namely: TD-SCDMA for China Mobile, WCDMA for China Unicom and CDMA2000 for China Telecom. Among them, China Mobile's technical standards are independently developed, and there are obvious disadvantages in many aspects. Contrary to the glory of the 2G era, China Mobile was strongly suppressed by China Unicom and Telecom in the entire 3G era.

In the 3G era, digital communications are moving towards data communications and data communications are no longer subsidiary to voice communications. Communication speed is greatly accelerated, with a minimum speed of 384Kbps, and through various technologies, it can reach 7.2Mbps, which is more than 30 times faster than 2G. The mobile Internet is starting to grow, and with bandwidth soaring, tariffs are getting lower and lower. In the 2G era, 1GB of traffic cost up to 10,000 RMB, but in the 3G era, the price of 1GB of traffic dropped to about 500 RMB. In addition to high quality calls, 3G mobile phones are also capable of multimedia communication and interoperable transmission with computers. The 3G network is in full bloom in China, with Apple and Samsung taking the lead in the mobile phone market with an overwhelming advantage, but the resulting surge in user numbers has laid a solid foundation for the launch of domestic smartphone brands in China, which will usher in an even faster 4G era.

4G: Data in Full Swing

As time draws out the ups and downs of everything, the communications industry continues to evolve in unexpected ways.

In the 1960s, Bell Labs invented orthogonal frequency division multiplexing (OFDM), and in the 1980s the framework for this technology was completed. In the early days OFDM was mainly used in military wireless high frequency communication systems, but due to its complex structure and the need for extensive digital signal processing, it was shelved due to the lack of mature hardware conditions, and in the 3G era it was Qualcomm's CDMA that dominated the market, so OFDM was almost unnoticed.

However, with the rapid development of digital signal processing hardware and integrated digital circuits, and the increasing demand for higher speeds in wireless communications, OFDM has finally seen the light of day and has come of age in an environment where both hardware and software are mature. After 1999, the Institute of Electrical and Electronics Engineers (IEEE) introduced the 802.11a Wi-Fi standard for wireless local area networks (WLANs), with OFDM as the physical layer standard and transmission speeds of up to 54Mbps, followed by the 802.11n, 802.11b, 802.16e and 802.11g Wi-Fi standards, with OFDM These standards, which are based on OFDM modulation and MIMO (Multiple

Input Output) technology, have greatly improved transmission speed, distance and spectrum efficiency, and have been a great success.

The 4G era is beginning to take shape, and the huge market is like a big cake, attracting countless businesses and attracting the IT industry. One IT company that has been instrumental in bringing OFDM back to the forefront of the telecommunications industry is Intel.

In 2005, Intel took the lead, along with Nokia and Motorola, in announcing the development of the 802.16 standard, calling it WiMax. The standard combines the long-abandoned OFDM and FDMA technologies into OFDMA, which is the core technology of 802.16 and has attracted a lot of attention from mobile giants, thus making OFDM rapidly popular. Compared to CDMA, OFDM is more simplified and can also effectively eliminate multipath interference. In 2009, 3GPP proposed Long Term Evolution (LTE), and in 2011 it proposed an upgraded version (LTE-Advanced), which planned to use OFDM and replace WCDMA. The major operators have also adopted LTE-Advanced, announcing the arrival of the fourth generation of communication standards. In view of the increasingly competitive situation, Qualcomm then integrated OFDM and MIMO to launch the UMB (Ultra Mobile Broadband) standard in an attempt to save the day and continue the glory of CDMA in the 3G era. At the same time, Intel's strong launch of WiMax is also a big thunder but little rain. Intel is an IT company, but came to grab the meat of the telecommunications industry, while WiMax evolved from Wi-Fi, the subordinate relationship is unclear, in the case of WCDMA base stations have been fully covered, LTE can be compatible with it, want to achieve marketability, but also have to build base stations from scratch. Furthermore, WiMax does not allow for soft switching, which leads to serious congestion and poor user experience when shared by a large number of users. Although WiMax was also supported by operators, it did not work well commercially and was declared a failure in 2010. With no tablecloths, no knives and forks, and a near-hopeless attempt to grab the meat, Intel threw in the towel.

By the fourth generation of mobile communications, China proposed TD-LTE (also known as LTE-TDD) based on TD, while Europe proposed FDD-LTE based on the original WCDMA. Both appear under the same keyword, LTE, which together with WiMax and 3GPP2's Ultra Mobile Broadband (UMB) technology is often referred to as 4G. In contrast to WiMax fixed wireless network technology, LTE uses orthogonal frequency division multiplexing (OFDM) for signal transmission, as well as Viterbi and Turbo accelerators. However, WiMax is a technology derived from IP, while LTE is derived from GSM/UMTS mobile radio technology. 3GPP plans to use OFDMA for the downlink and SC-FDMA (Single Carrier FDMA, also known as "DFT Extended OFDM") for the uplink of LTE, which will reduce the power consumption of mobile phones. LTE systems are able to use different band widths depending on the available spectrum, making LTE more advanced than WiMax in terms of mobility. FDD and TDD are two modes, with the former being used for paired spectrum and the latter for unpaired spectrum.

4G, which combines 3G and WLAN, marks the full explosion of the data age: The unprecedented speed allows audio, video and images to be transmitted quickly and downloaded at speeds of over 100 Mbps, meeting the needs of virtually all users of wireless network services and expanding the range of deployment dramatically, giving an overwhelming advantage over the mobile communications of the past.

In August 2013, the State Council held an executive meeting where Premier Li Keqiang specifically proposed to speed up the issuance of 4G licences for deployment with TD-LTE. The 4G network has been steadily expanding in China, forming lines and lines into pieces. As TD-LTE technology flexibly supports 1.4, 3, 5, 10, 15 and 20MHz bandwidth and uses OFDMA for downlink, the maximum speed reaches 100Mbps, which can meet the requirements of high-speed data transmission, bringing users a far better experience than expected, and the number of users has risen sharply. By the end of June 2018, the number of 4G subscribers in China exceeded 1.11 billion. With the booming development of 4G networks, basic telecoms companies have accelerated the construction of mobile networks, with the total number of base stations of the networks of China's three major telecoms operators now exceeding 6.4 million, with over 3.5 million base stations for 4G, far exceeding the total number of 4G base stations of other countries in the world. It is on this rapidly expanding network that China's mobile phone industry has also seen a major turnaround.

With the arrival of 4G, mankind has truly entered the era of mobile Internet with the increase in Internet speed and network coverage. A large number of video-based services began to explode, with video broadcasting services shifting from traditional television to the Internet, with on-demand services becoming the mainstay of many Internet video services and fees becoming mainstream, with users getting used to the membership model.

The emergence of live streaming has largely influenced people's mode of entertainment and communication. In the process of live broadcasting, a large number of rewards become the main source of income for the platform and the anchors. The traditional free-to-use model of the Internet is fading, and charging for services, either through in-app fees or rewards, is becoming more widely accepted.

In this mobile Internet system, the terminals are gradually shifting from PCs to smartphones. Apple has rewritten the user experience with iOS, tiled desktop interaction and touch screen, and with Apple's impetus, Android has also introduced tiled desktop interaction and touch screen into smartphones, gradually forming two ecosystems, iOS and Android. With thousands of apps integrated through the app shop, most of the services in social life can be done through apps.

The earliest mobile Internet services began to emerge in the 3G era, mainly in the US, and many smartphone-based services such as Twitter and Facebook have completely disrupted the traditional Internet. These businesses were soon copied and learned by Chinese Internet developers, who went on to develop products such as Weibo and WeChat.

The arrival of 4G, with its large bandwidth and strong coverage, China's excellent network coverage and increasingly low Internet access costs, has driven the development of mobile Internet in China. Through the accumulation and learning of 3G, China has surpassed the United States in the 4G era and has become the most active and well-established country in the world in this field.

The most important feature of the Chinese mobile Internet is the integration of all services through social networking, the most representative of which is WeChat. Today, WeChat has become a powerful service platform that integrates a wide range of services such as mobile gaming, mobile payments and transportation services, and through the power of social networking, these services are being rapidly promoted and the operators involved are reaping good financial rewards.

Another major feature of China's mobile Internet is the ability to pay electronically in every corner of society, with two platforms—Alipay and WeChat Pay—making payment extremely easy. It is because of the default presence of high quality 4G networks in every corner that people can go out without cash. Today, everything that needs to be paid for, from ordinary life to public services, can be done by electronic payments.

Because smartphones provide location capabilities and mobile electronic payments provide powerful payment capabilities, shared services in China are growing rapidly, with bike-sharing and car-sharing services growing rapidly and services like takeaway penetrating into everyday life. Hundreds of millions of orders a day make social life extremely convenient. 4G has allowed data services to explode and China has truly entered the mobile Internet era. This era not only provides high-speed information transmission, but also makes many services in life mobile and intelligent through positioning, mobile terminals and mobile electronic payments. In the process, people enjoy the convenience and efficiency of social life. Today, from plane check-in to train ticket booking, to bus, subway, in China, people can do all these operations with a smartphone. The phenomenon of queuing all night for a ticket has gradually disappeared.

The most fundamental basis for the rapid development of mobile e-commerce, mobile payment and shared services is the high speed, full coverage 4G network and relatively inexpensive communication tariffs, which are the cornerstones of the mobile Internet business explosion.

5G: Mankind Will Have a Smart Internet

If 4G has changed people's lives, the arrival of 5G will change our society, that is to say, this new change will be much more profound in both breadth and depth.

Examples of how 4G has changed lives can now be seen everywhere, such as mobile payments, bike-sharing and m-commerce, all of which would have been difficult to achieve before 4G. If someone had said that everyone could go out with their mobile phone and do a lot of things, they would have thought it was a fantasy, but in the 4G era it has become commonplace.

At the same time, 4G has enabled society to cross the digital divide. Take m-commerce as an example. Before the 4G era, it was difficult to get old women in remote areas to use computers to access the Internet and sell their crops, such as sweet potatoes, to the cities because the threshold for learning to use computers was very high. But in the 4G era, smartphones have helped people cross the digital divide and e-commerce is now possible for people in remote areas.

In hindsight, there is nothing special about mobile payments, and it even feels like it should be. But the fact is that without 4G, these features would not be possible at all.

In the 4G era, China is leading the world in many aspects, and has played a very important role in economic development, changing people's lives, improving social efficiency and reducing social costs.

In the 5G era, mankind will enter an era of smart Internet that integrates mobile connectivity, smart sensing, big data and smart learning. In the 5G era, the ability of mobile connectivity breaks through traditional bandwidth limitations, while time delays and the ability to access a large number of terminals are fundamentally solved, fundamentally breaking through the ability to transmit information, enabling intelligent sensing, big data and intelligent learning capabilities to be brought into full play, and integrating these capabilities to form a powerful service system.

This system of services will not only transform society, but will also permeate the field of social management and change all aspects of life.

One of the most important aspects of 5G's ability to transform society is the ability to build an efficient social system at low cost. For example, air quality is a hot topic of great concern to people today, but the monitoring system established by traditional technology is costly and inefficient, and cannot be truly comprehensive. There are also only 35 air quality monitoring points in Beijing, making it difficult to monitor pollution sources effectively. Through the low-power network of 5G, a large number of monitoring devices will be built, turning street lights and utility poles into monitoring points, which will not only provide an accurate understanding of air quality conditions, but will also provide a more scientific basis for controlling corporate emissions and understanding the causes of pollution.

Some people may ask, are many capabilities still achievable through 4G or even 2G networks? The answer is yes, but relying on traditional networks is not only costly, it is also impossible to support the large number of devices that can be accessed. It is the Internet of Everything capability of 5G that will truly support this mass access.

As a public network, 5G will be sliced into multiple slices to provide a wide range of services in the areas of smart transportation, smart home, smart health management, industrial Internet, smart agriculture, smart logistics and social services, which will not only enhance social life and make people's lives more convenient, but also improve social management capabilities, making social management more efficient and improving social public services in general.

The value of 5G is not only faster speeds, but also low power consumption, low latency, and the interconnection of everything, which allow the network to extend its capabilities significantly. With the advent of 5G, the world will not be the same as it was in the past.

The Three Major Scenarios for 5G

When it comes to 5G, it's all about scenarios—where to use it. In this regard, the International Organization for Standardization 3GPP has defined three main scenarios: eMBB (Enhanced Mobile Broadband)—High volume enhanced mobile broadband services such as 3D/Ultra HD video; mMTC (Massive Machine Type Communication)—Large Scale IoT Services; uRLLC (Ultra-Reliable Low-Latency Communications)—services such as driverless and industrial automation that require low latency and highly reliable connections.

eMBB: Enhanced Mobile Broadband is a significant increase in user experience speed based on existing mobile broadband service scenarios. Today we use 4G network, the actual experience of the average user speed 6Mbps upload, download 50Mbps, this speed is far from meet the needs of users, the experience is not good enough, especially for some large traffic requirements of high business, such as live video, and so on. The 4G live video upload speed is only about 6Mbps, which cannot provide high-definition video, and even this speed cannot be guaranteed in some crowded places. The value of enhanced mobile broadband is to increase the speed of the original mobile broadband significantly, reaching a theoretical 1Gbps or so, the user experience will change dramatically.

The importance of enhanced mobile broadband for a wide range of bandwidth-hungry services such as live streaming, HD video, HD video retransmission, VR experiences and more cannot be overstated. In countries such as the US and Germany, where the deployment of fibre optic cables is poor, there is still a degree of limited access to the Internet. The use of enhanced mobile broadband can compensate to some extent for the lack of fibre optic cables and enhance the user experience of broadband access. eMBB can be deployed in standalone or non-standalone configurations: The main network is 4G, but enhanced mobile broadband is being deployed in key areas.

eMTC: Large-scale Internet of Things, enabling massive machine-like communications. One of the main values of 5G is that it breaks through human-to-human communication and makes human-to-machine and machine-to-machine communication possible. A large number of IoT applications require communication, and there are two basic requirements for communication in IoT applications: low power consumption and massive access.

A large number of IoT applications such as utility poles, parking spaces, manhole covers, home door locks, air purifiers, heaters, refrigerators, washing machines, etc. have to be connected to the network. A significant number of IoTs cannot be powered by a fixed power source and have to use batteries, which means that it is very difficult to deploy if the communication part requires a large power consumption, which will greatly limit the development of IoT. eMTC offers the ability to reduce power consumption to extremely low levels, allowing large numbers of IoT devices to be deployed easily without needing to be recharged for a month or more.

The addition of a large number of IoT applications brings with it another problem: the number of application terminals will increase dramatically. It is expected that by 2025, the number of mobile terminals in China will reach 10 billion, including more than 8 billion IoT terminals, which requires the network to have the ability to support a large number of devices to access, and the current 4G network obviously does not have the ability to support such a huge number of accesses, eMTC will provide low-power, massive access capabilities to support a large number of IoT devices to access.

uRLLC: Ultra high reliability ultra low latency communication. In traditional communication, the requirements for reliability are relatively low, but unmanned vehicles, industrial robots and flexible intelligent production lines place higher demands on communication, which must be highly reliable and with low latency.

High reliability means that the network must remain stable, ensuring that it will not be congested, interfered with or constantly affected by external influences during operation. The best 4G network latency is 20 milliseconds, but uRLLC requires a latency of 1 to 10 milliseconds to provide highly stable and secure communication, so that unmanned vehicles and industrial robots can respond to commands in the first instance and execute them quickly and in a timely manner. This requires a variety of technologies, such as edge computing and network slicing, to support more highly reliable communication scenarios.

These three scenarios basically represent the basic vision of the world mobile industry for 5G.

Six Essential Features of 5G

The three main scenarios of 5G will not only address the speed that has been a concern, allowing users to get faster speeds when using communications, but also put forward higher requirements for power consumption, latency, etc. Some aspects are completely beyond people's understanding of traditional communications and more capabilities have to be integrated into 5G. In these three scenarios, 5G also has features that are completely different from traditional mobile communications, some of which are not included in the three scenarios, but must be gradually improved to become features of the 5G system. 5G has six essential features.

High Speed

With each new generation of mobile technology, the most direct experience for users is the increase in speed.

When the 3G era first arrived, people were pleasantly surprised, but a few years later, the growing demand could no longer be met by 3G, so people began to look forward to 4G. With the advent of the 4G era, there has been a major breakthrough in Internet speed and people are amazed that they can transfer files and watch videos on their mobile phones without getting stuck, and download an HD movie in just a few minutes. With speeds of up to 1Gbps and as fast as 10Gbps, 5G is no longer calculated in Mb, and it takes only a few seconds to download an ultra-clear movie, or even less than a second, which is as fast as a rocket! This breathtakingly high speed will be fully applied to all smart technology mobile terminals in the 5G era.

The significant increase in Internet speed will ensure the quality of our Internet experience. The first online content was called newsgroups, no images, only text content. A friend sent me a highly compressed greetings video for New Year's Eve, only 2M, but it took me hours to download it. In the 3G era, when we use features such as Weibo, pictures are defaulted to thumbnails, so when you want to see them, you need to click on them to open them, in the 4G era, these images are opened by default, which is the result of a significant increase in network speed.

In the 5G era, it's not just mobile phones that deserve our attention. High-speed 5G networks will host enhanced mobile broadband (eMBB) application scenarios, the closest to everyday life is watching ultra-high-definition video at home on a smart TV. At the same time, a wide range of terminal products are being actively developed to meet the high traffic applications made possible by the ultra-high speeds brought about by the 5G era.

The average 4G user can experience speeds of 6Mbps for uploads and 50Mbps for downloads, and around 150Mbps through carrier aggregation technology. 5G can theoretically achieve a speed of 20Gbps per base station, and the actual effectiveness of each user may be close to 1Gbps. Such a high speed is not only as simple as a user downloading an ultra-HD movie in one second, but it will also revolutionise a large number of services and applications.

In the traditional Internet and 3G era, traffic was a very precious resource due to the speed of the network, and all social software was an access mechanism, which meant that users had to be online in order to receive data. In the 4G era, network speeds have increased, bandwidth is no longer a precious resource, and social applications have become push mechanisms, where all information can be pushed to your phone and you can see it at any time, this means that your phone is always online, a change that has made a world of difference to the user experience and has seen an explosion of users.

The greatly increased speed of 5G will also have a huge impact on related services, not only making traditional video services a better experience, but also giving rise to a large number of new market opportunities and operational mechanisms.

Take a very typical example. The live streaming business has seen phenomenal growth in the 4G era, bringing huge business opportunities, but the upload speed of 4G is only 6Mbps, and when more people are using it at the same time, this speed is not guaranteed, lagging is common, and the live streaming effect is affected, especially for some content that needs to support HD live streaming, which is a poor experience. With 5G upload speeds of around 100Mbps, network slicing technology can also ensure that certain users are not affected by congestion and that live broadcasts will be better. In this context, every user has the potential to become a live TV station, and the current explosion of new media and traditional live TV programmes is bound to face a whole new level of competition.

High speeds also lead to new business opportunities. Virtual Reality (VR) could then be a breakthrough with 5G. One of the main reasons why the VR experience is so poor today is that it is not supported by speed. VR requires a bandwidth of 150Mbps or more for good HD transmission, which is not possible on most networks. With the arrival of 5G, the VR experience will be greatly improved and the development of the VR industry is fully expected.

The high speed will also support the move from concept to practical applications such as telemedicine and tele-education. The basis for the viability of telemedicine is low cost, but also the need for high-definition image transmission and the need for low latency operation, all based on a high-speed network.

High speed is one of the most distinctive features that sets 5G apart from 4G. Humans' pursuit of speed is endless, so there will never be enough. 3G is not enough, 4G is not enough, and 5G is not enough. Humanity will always seek to support greater bandwidth, higher speeds and, on top of that, more services with a variety of new technologies to make traditional services and services a better experience.

Ubiquitous Network

The huge growth of various services places greater demands on the 5G network, which needs to be all-encompassing and widespread. Only a ubiquitous network can support increasingly rich services and complex scenarios. At present, for example, we can tolerate the absence of a network in underground car parks, although there is a certain amount of hassle. If driverless cars are widely adopted and there is still no network in underground car parks, these driverless cars will not be able to automatically enter the garage to park and charge, so this network must be widely available.

There are two dimensions to the ubiquitous network: one is broad coverage and the other is deep coverage. Extensive coverage means that the human footprint extends to places that need to be covered, such as mountains and canyons, where people rarely go before and do not necessarily need network coverage, but in the 5G era, these places must have a network presence, because whether it is intelligent transportation or other services, they need to be managed through a stable and reliable network, which cannot be managed in places where there is no network. At the same time, by covering the 5G network, sensors can be deployed in large numbers to monitor the natural environment, air quality, changes in the landscape of mountains and rivers and even earthquakes, and 5G can provide a network for more of these applications.

Deep coverage is where people already have a network deployed in their lives, but need to move into higher quality deep coverage. Today, most people have 4G networks in their homes, but the quality of the network in small spaces such as bathrooms is often not so good, and most underground garages have no signal, so if you want to deal with things in this environment, you will face the embarrassment of no network. In the 5G era, special places such as bathrooms and underground garages with no signal, where the network quality was previously poor, can and need to be covered by a high-quality network. Because in the future, the flush toilet in your home may need to be connected to the Internet, the toilet may automatically help you to do urine routine check and send it to the cloud, through the big data comparison, to determine your health situation, through various aspects of improvement to improve your physical condition, this may become an important component of the intelligent health management system, and thus play a very important role in the improvement of human body.

In some ways, ubiquitous networks are more important than high speeds. Imagine building a network that covers only a few places with high speed, but does not guarantee a large area of 5G services and experience, equivalent to a quality product that only a very small number of people can experience, which is certainly not possible. Ubiquitous networks are not mentioned in the three major 3GPP scenarios, but the requirement for ubiquity is implicit in all scenarios.

Before the 4G era, we often had problems with no or weak mobile phone signals, especially in remote areas. This is because in the 3G and 4G era, we used macro base stations. Macro base stations are very powerful, but they are also relatively large, so they cannot be deployed densely, resulting in strong signals close to them and weaker signals the further away they are. But in the 5G era, micro-base stations will be built gradually, and there will be little need to worry about signal shortages.

Micro base stations, i.e. small base stations. The deployment of micro-base stations can fill the gap of macro-base stations, covering the end-point communication that macro-base stations cannot reach, providing the possibility of full realisation of ubiquitous network, enabling all intelligent terminals to break through the restrictions of time, place and space, and connect to the network signal in any corner.

Low Power Consumption

As technology continues to evolve, network speeds are becoming faster and faster, while device power consumption is becoming higher and higher. A large part of the reason why Google Glass is not commercially available is that it consumes too much power and the user experience is too poor.

From this perspective, reducing power consumption is a big issue, and 5G must consider the power requirements to support large-scale IoT applications. A practical example is that wearable products have made some progress in recent years, but they have also encountered many bottlenecks, among which the poor experience is the main reason why it is difficult to enter the life of the general public. Smart watches, for example, need to be charged every day or even for several hours, resulting in a very poor user experience. In the future, IoT products need communication and energy, although today you can achieve communication through a variety of means, but most of the energy supply can only rely on the battery, in order to ensure the use of products for a long time, the power consumption must be brought down, so that most IoT products are charged once a week, or once a month to improve the user experience, more and more rich IoT products will be widely accepted by the general public.

Low power consumption is mainly achieved by two technologies, eMTC led by Qualcomm and others in the US and NB-IoT led by Huawei.

eMTC is based on the evolution of the LTE protocol, which has been trimmed and optimised in order to make it more suitable for object-to-object communication and also to make it cheaper. eMTC is deployed on cellular networks and its user devices can directly access existing LTE networks by supporting 1.4MHz of RF and baseband bandwidth. eMTC supports peak rates of up to 1Mbps upstream and downstream.

The construction of NB-IoT is based on the cellular network, which only consumes about 180KHz of bandwidth, and can be directly deployed in the GSM network, UMTS network or LTE network to reduce deployment costs and achieve smooth upgrades.

NB-IoT does not require rebuilding the network as much as the core technology of 5G. Although the transmission speed of NB-IoT is only about 20K, it can significantly reduce power consumption, allowing devices to go a long time without changing batteries. This feature is beneficial for the large-scale deployment of various devices and meets the low-power requirements of 5G for IoT application scenarios. NB-IoT, like eMTC, is an integral part of the 5G network architecture.

The requirements for low power consumption are very extensive, to give a typical example. For river water quality monitoring, a few dozen kilometres or a few kilometres to set up a monitoring point, monitoring results are not accurate enough, to find the source of pollution is very difficult, and the establishment of a large number of conventional monitoring points, the cost is too high, which requires the establishment of a large number of low-cost monitoring points, timely return of data. For river water quality monitoring, a few dozen kilometres or a few kilometres to set up a monitoring point, monitoring results are not accurate enough, to find the source of pollution is very difficult, and the establishment of a large number of conventional monitoring points, the cost is too high, which requires the establishment of a large number of low-cost monitoring points, timely return of data. If low-power technology is used and the monitors are placed along the river and the batteries are changed once every six months, the cost of maintenance is very low, thus creating a valuable application.

Low Latency

A new scenario for 5G is highly reliable connectivity for driverless, industrial automation. Under normal circumstances, a time delay of 140 milliseconds is acceptable for human-to-human communication and will not affect the effectiveness of the communication. However, for scenarios such as driverless and industrial automation, such latency is unacceptable. The ultimate requirement for latency in 5G is 1 millisecond or less, which is very demanding, but necessary. 3G network latency is about 100 milliseconds, 4G network latency is about 20–80 milliseconds, and in the 5G era, the latency will gradually drop to 1–10 milliseconds.

The low latency characteristics of 5G will certainly lead to a major explosion in the fields of autonomous driving and connected cars. Generally speaking, driverless cars need to be interconnected with the central control centre and the car, and the car should also be interconnected with each other. In the event of braking at high speed, the information needs to be transmitted to the car instantly, and the car's braking system will react quickly. In 100 milliseconds or so, the car will rush forward a few metres, which requires a minimum time delay to transmit the information to the car and get a timely response, otherwise the consequences are unthinkable.

In addition, a large number of sensors and cameras will be deployed on the relevant traffic hubs to capture video and form dynamic flow maps through big data transmission, so that pedestrians can easily and intuitively see the actual traffic situation, and the intelligent traffic system will be fully completed. The technology is also expected to be used for live broadcast of matches, allowing people to see personalised live sporting events on their mobile devices and watch them from all angles.

The application of intelligent transportation is just the beginning of the 5G era, and the real king is the drone.

In unmanned aircraft, the requirements for low latency are also high. For example, when hundreds of unmanned aircraft are flying in formation, the distance and movements between each aircraft must be extremely precise in order to ensure safety, and even a delay in the transmission of information by one of them for too long could lead to a major catastrophic accident. In the field of industrial automation, the robot arm also requires ultra-low latency in order to operate the assembly of parts with a high degree of refinement and to produce high-quality products.

Currently, the requirements are not as high when it comes to traditional human-to-human communication, or even human-to-machine communication, as humans are relatively slow to react and do not require as much efficiency and refinement as machines. To meet the low latency requirements, ways to reduce latency must be found in the construction of 5G networks. It is foreseeable that technologies such as edge computing will be used in the network architecture of 5G.

In 2017, China's first low-altitude digital application innovation base was inaugurated in Shanghai, which will build a 4G+5G network, conduct low-altitude flight experiments and explore the possibilities of "air corridors". With the greater bandwidth, higher speeds and ultra-low latency of 5G, drones will be able to achieve more precise control and timely communication.

Another important application area for low latency is industrial control. This area has the highest requirements for the time delay, a high speed CNC machine tool, the order to stop, this information if not delivered in time, but a high time delay, it is impossible to ensure that the production of parts is high precision. Low latency means that the machine reacts as soon as the information is delivered, so that precision can be guaranteed.

Low latency requires a large number of technologies to work together, combining edge computing and other technologies with traditional networks to provide special services and guarantees for specific areas.

The Internet of Everything

The basic means of contact for mobile communication is cellular communication, now a base station can only connect basically 400 or 500 mobile phones, the International Telecommunication Union expects one million terminals per square kilometre. Ericsson has a prediction that there will be 50 billion connections for humanity in the future. Our current forecast is that by 2025, there will be 10 billion mobile terminals in China.

In traditional communications, the terminals are very limited, because in the era of fixed-line telephony, telephony was defined by groups of people, for example, one phone at home and one phone in the office. In the age of the mobile phone, the number of terminals has exploded, because the mobile phone is defined by the individual. Smart terminal products in the 4G era are no longer limited to mobile phones, but smart bracelets, tablets, smart appliances and drones have all entered people's daily lives.

Why do you think there will be 10 billion mobile terminals in China by 2025? There are two reasons.

The first reason is that in the 5G era, terminals will no longer be defined by people, but each person may have several terminals, and each household may have several terminals. There will be an even greater variety of smart products, which will be interconnected through the network, creating a truly smart world of things. In the future, people will no longer have the notion of being connected to the Internet, it will become the norm.

Data shows that in 2018, China's mobile terminal users have reached 1.5 billion, with mobile phones dominating. From the point of view of the development trend, the terminal in the 5G era access network, no longer mainly mobile phones, will be extended to more products in daily life. In other words, glasses, pens, purses, belts, shoes, etc. could all be connected to the Internet and become smart products. Doors, windows, locks, air purifiers, humidifiers, air conditioners, refrigerators and washing machines can all be connected to the 5G network and transmit information to each other, so that ordinary families can become fully intelligent smart families.

The second reason is that devices in social life that were previously unlikely to be connected will also be connected and become more intelligent. For example, cars, manhole covers, utility poles and rubbish bins are all public facilities that previously had a single function and were not very intelligent. 5G will give these devices new capabilities and make them smart devices. 4G is changing lives, 5G is changing society, and herein lies the essence.

Big data used to be accessed through traditional channels, but if we have 10 billion terminals that can transmit data in the future, the amount of data we have will increase dramatically. With such a large amount of data, the ability of artificial intelligence becomes powerful and valuable, which is very valuable for the efficiency of society.

The 5G era will also create many new products that were not previously available in people's daily lives, such as products such as the Environment Power at home, which will be used to monitor indoor air quality and intelligently control the air purifier, air conditioner and even the heating at home based on the monitoring results. Toilets will also be smarter, not only for flushing but also for daily health checks.

We can imagine a future scenario:

When you get home from work, the room's intelligent air conditioning and lights automatically turn on according to your outdoor body temperature, and the temperature is adjusted to a comfortable indoor temperature; The smart microwave oven automatically heats up dinner by induction. After a meal, the bath is automatically drained and adjusted to the right temperature for you; the smart electric toothbrush records your oral health. If you need to travel at this time, the garage door opens automatically and as soon as you get in the car you can immediately see the real-time road conditions to your destination.

In the future, all facilities, even wearable products, may be connected to the mobile network, forming an immensely powerful database that will seamlessly connect the virtual with the real, bringing about a new era of intelligence.

The Internet of Everything has also brought about a market explosion. With a large number of smart hardware entering the 5G network, the number of connected devices will increase from a few hundred million or a billion to ten billion. A large number of devices can become information-gathering terminals for big data, thus significantly improving service capabilities. It is on this basis that cloud and artificial intelligence can be developed more extensively.

Restructuring Security Systems

Security does not seem to be a fundamental issue in 3GPP discussions, but in my personal opinion it should also be a fundamental feature of 5G. Specifically, the traditional Internet is to solve the problem of information transmission speed and barrier-free transmission, freedom, openness and sharing are the basic spirit of the Internet, but the intelligent Internet built on 5G is more diversified, in addition to the basic functions of the traditional Internet, but also to establish a new system of society and life. This is why the spirit of the Smart Internet has become one of security, management, efficiency and convenience.

Security is one of the first requirements for the Smart Internet in the 5G era. If there is no security guarantee, 5G cannot be built; after 5G is built, if the security system cannot be rebuilt, it will have huge destructive power.

It can be imagined that if the self-driving system is easily broken, like the plot of the movie, the cars on the road may be easily controlled by hackers. If a smart health system is breached, a large number of users' health information is compromised. What would the world look like if the smart home was compromised …? These horrific scenarios shouldn't happen, and security is not something that can be tinkered with.

In the construction of 5G networks, security issues should be addressed at the bottom. Security mechanisms should be incorporated from the beginning of the network, information should be encrypted, the network should not be open, and special security mechanisms need to be set up for special services. The Internet is not completely neutral and fair. Take a simple example. In terms of network assurance, users may face congestion if there is only one system to ensure smooth network access for the average user. However, in the field of intelligent transportation, multiple systems are required to ensure its safe operation and network quality, and in case of network congestion, the network of the intelligent transportation system must be ensured to be smooth, and this system cannot be accessed and managed and controlled by general terminals.

According to national e-commerce transaction statistics, the proportion of mobile phone payment users has increased from 33% in 2014 to over

75% today. In the future, in the 5G era of the Internet of Things, every product will be equipped with sensors, making it possible to make unattended and automatic purchases. Currently, unmanned supermarkets are being piloted in some cities across the country.

In the current mobile payment system, security is also a major concern, with card theft, payment fraud and illegal financing commonplace, which has become the main reason why many users are reluctant to choose mobile payments. In the 5G era, with the technology of big data, cloud computing and artificial intelligence, the security of mobile payment will be gradually solved. The major financial institutions have been keeping pace with the times, developing new smart payment products, smart terminals are expected to become mobile financial security terminals, and the new security system will be reconstructed.

With the advent of 5G, the traditional Internet TCP/IP protocol will also be put to the test. The security mechanisms of the traditional Internet are very weak and information is transmitted directly without encryption, a situation that cannot continue in the age of smart Internet. With the large-scale deployment of 5G, more security issues will emerge, and countries around the world should form a new mechanism on security issues, and finally establish a new security system.

The Core Technologies for 5G

With the help of the Internet, global economic integration has accelerated and "the world is becoming flat" has become a consensus. This sweeping convergence force is also becoming apparent in the field of mobile technology. It is like some kind of viscous sap that not only gives rise to a structural change in the mobile communications system, but also permeates every aspect of the way technology is implemented.

5G is not a single technology, but a comprehensive system formed by a large number of technologies, which will be continuously improved during the construction of 5G. In the meantime, new technologies will emerge and continue to be refined. This book is not a technical monograph and the discussion of the technology is superficial and can be skipped for the specialist reader. The description of the technology is only a basic introduction for the general reader.

The six basic features of 5G, including high speed, ubiquitous network, low power consumption and low latency, guarantee the basic user experience in the 5G era, while the core technologies provide the guarantee for the realisation of the six features, which are the effective means to protect the new mobile communication era.

In summary, the core 5G technology is built around three main objectives, and will continue to evolve in a more intelligent and versatile direction, building on the technological heritage of the past. The three main objectives are:

Firstly, activating the stock of online resources.
Secondly, to tap into the incremental resources of the network (new frequency resources).
Thirdly, flexible combinations for diverse network resource allocation (introduction of new architectures).

Ultra-Dense Heterogeneous Networks

In the future, 5G will develop towards high speed, ubiquitous networks, etc. The vision of the Internet of Everything is one million devices in an area of one square kilometre, so in future 5G networks, reducing the cell radius and increasing the number of low-power nodes is one of the core technologies that will ensure that future 5G networks support a 1000-fold increase in traffic, which means a particularly dense network. At the same time, in order to meet the requirements of the ubiquitous network, there will certainly be a large number of base stations in the future. In the 2G era there were only tens of thousands of base stations, in the 3G era there were hundreds of thousands of base stations, in the 4G era there were more than 5 million base stations and in the 5G era 10–20 million base stations are possible.

Why are there so many base stations? In the past, communication methods used to be based on the lower frequency bands of the spectrum, which have a relatively high penetration capacity—the lower the frequency, the higher the penetration capacity.

In order to make the bandwidth of 5G very wide, the bypass capability is very poor. 5G uses frequencies from 28GHz to 32GHz, that is, millimetre waves, which basically have no penetrating ability. Radar uses millimetre waves, which are reflected back because they cannot penetrate the aircraft.

The use of millimetre wave frequencies for communications means that there is no way to penetrate barriers, so many micro-base stations are needed for dense deployment. The densely deployed network brings the terminals closer to the nodes, resulting in a significant increase in network power and spectrum efficiency, while also extending network coverage, expanding system capacity, and increasing the flexibility of services across different access technologies and coverage levels.[1]

What is described above is hyperdense. So, what does it mean to be heterogeneous?

Heterogeneity means different structures. Although ultra-dense heterogeneous network architectures hold great promise for 5G, as the number of nodes increases massively and the distance between nodes decreases, network deployments will become increasingly dense and network topologies will become more complex, making them prone to incompatibility with existing mobile communication systems. In 5G communication network, interference is a problem that must be solved.

5G networks require a range of measures to guarantee system performance, mainly the implementation of different services in the network, the coordination scheme between various nodes, the choice of network, and energy-efficient configuration methods. This way of organising multiple networks into a system is called hyperdense heterogeneity.

Ultra-dense heterogeneous network technology is an inevitable product of the development of mobile communications to the stage of convergence. As the future of mobile communications continues to enrich application scenarios, the requirements for network information transmission

[1] Zhao GF, Chen J, Han YB, Xu C. A review of key technologies for 5G mobile communication networks [J]. Journal of Chongqing University of Posts and Telecommunications (Natural Science Edition), 2015, 27(4):441–452.

will be non-uniform over time and location. The previous macro-cell-based mobile network architecture, which was designed for area coverage, is no longer sufficient to meet the exponentially growing segmentation needs.

Ultra-dense heterogeneous network technology has emerged as an innovative way of tackling this challenge head-on. It goes beyond the scope of operators and technical systems, and integrates different networks together, providing a strong guarantee for the large capacity, diversity and flexibility of network systems in the 5G era.

While ultra-dense heterogeneous network architectures hold great promise for the 5G era, they also bring with them a number of entirely new issues.

Firstly, there is the issue of compatibility. The reduced distance between nodes and the deployment of ultra-dense networks will make the network topology more complex and increase the probability of incompatibility with existing mobile communication systems.

Secondly, there is the issue of interference. In 5G mobile communication networks, the main types of interference are co-channel interference, interference from shared spectrum resources and interference between different coverage levels. The coexistence of multiple wireless access technologies and coverage levels is an important issue that needs to be studied in depth to address the performance impairment caused by these interferences. In addition, the interference coordination algorithms of existing communication systems can only solve the problem of individual interference sources. In 5G networks, the transmission loss of neighbouring nodes generally does not differ significantly. Multiple sources of interference of similar strength will further deteriorate network performance and make it difficult for existing coordination algorithms to cope.

Finally, there is the issue of network switching. In ultra-dense networks, many network nodes rely on user deployments, which are randomly switched on and off, and the network topology and interference changes continuously and dynamically. We need new switching algorithms and network dynamic deployment techniques to meet the mobility needs of our users.

Self-Organising Networks

In traditional mobile networks, network deployment and operation and maintenance are mainly done manually, which consumes a lot of human resources and increases operating costs, and network optimisation is not ideal.[2] In the 5G era, the existing mobile network will face many new challenges, such as how to deploy, operate and maintain the network, mainly due to the complex relationship between the various wireless access technologies and the varying coverage capabilities of the network nodes.

A simple example. If a part of a network system has to be divided for intelligent traffic use, and intelligent traffic services have relatively high requirements for the quality of the network, so the meaning of self-organising network (SON) is that the network should be organised according to different services in the process of definition, that is, for a variety of different requirements of the network can be built through a self-organising system, in the large network system to provide special services for certain users. Therefore, the intelligence of self-organising networks will become a key technology essential for 5G networks.

The key problems solved by self-organising network technology mainly include: (1) Self-planning and self-configuration in the network deployment phase; (2) Self-optimisation and self-healing in the network maintenance phase.[3] The purpose of self-planning is to dynamically plan and execute the network while meeting the needs of the system in terms of capacity expansion, service monitoring or optimisation of results. Self-configuration, i.e. the configuration of new network nodes is plug-and-play, with the advantages of low cost and easy installation. The purpose of self-optimisation is to reduce business workload and achieve the effect of improving network quality and performance.[4] As for self-healing, as the name suggests, it means building a network system that can automatically identify and locate problems and troubleshoot them at the same time,

[2] IMT-2020(5G) Promotion Group.5G Vision and Requirements, white paper [EB/OL]. [2014-05-28]. http://www.IMT-2020.cn.

[3] He Jing, Chang Jiang. Self-organizing network (SON) technology and the evolution of standardization [J]. Post and Telecommunications Design Technology, 2012(12):4–7.

[4] Hu Po, Li Wenyu, Song Aihui. Advances in self-organizing network technologies and standards [J]. Telecommunications Network Technology, 2012(12):53–57.

significantly reducing maintenance costs and avoiding any impact on network quality and user experience.

Content Distribution Networks

In the 5G era, with the dramatic growth of audio, video and image services and the continued expansion of subscribers, the strong market demand will naturally lead to an explosion in network traffic, which will affect the quality of service for users accessing the Internet.

If there has been a particularly popular TV show recently and everyone is accessing a particular server together, it can cause network congestion. In the 5G era, how to carry out efficient content distribution, especially for high traffic business content, how to do to reduce the user access to information delay, become a major problem that network operators and content providers must solve.

Efficient content distribution is not completely solved by increasing bandwidth, as it is also influenced and constrained by multiple factors such as blockages and delays in transmission routing and the processing power of the web server, which are closely related to the distance between user servers. The content distribution network (CDN) will play an important role in supporting the capacity and user access of future 5G networks.[5]

A content distribution network is a new layer added to the traditional network, namely the intelligent virtual network. Using big data analysis, the CDN system takes into account the connection status of each node, the load situation and the distance of the user, and distributes the relevant content to the CDN proxy servers close to the user, so that the user can get the information they need nearby.[6] If a lot of users nearby like to watch "It's All Good", the show will be stored on the network node here, making it possible to reduce network congestion, reduce response times and improve response times.

[5] Wang Wei. Study on the optimization method of CDN content distribution network [D]. Wuhan: Huazhong University of Science and Technology, 2009.

[6] Zhao GF, Chen J, Han YB, Xu C. A review of key technologies for 5G mobile communication networks [J]. Journal of Chongqing University of Posts and Telecommunications (Natural Science Edition), 2015, 27(4):441–452.

In the 5G era, with the rapid growth in the number of smart mobile terminals, the demand for mobile data services and the quality of service required by users is also increasing, and content distribution network technology can meet these needs, so it will become one of the key technologies necessary for 5G.

D2D Communication

D2D communication, device-to-device communication (D2D), is a direct data transmission technology based on cellular systems in close proximity. Currently, the standardisation organisation 3GPP has included D2D technology in the development framework of the new generation of mobile communication systems as one of the key technologies of the fifth generation of mobile communication. Cisco predicts that global mobile data traffic will be 10 times higher in 2019 than in 2014, and the number of devices connected to IP networks will reach 10 billion.[7] With the rapid growth of data traffic and the number of terminals accessing the network, the architecture and structure of communication networks are facing enormous challenges. In order to cope with the problems of network densification and differentiation, the central device of any network or communication system cannot be expected to command and dispatch the behaviour of end nodes on a large scale and with high efficiency. In order to cope with the problems of network densification and differentiation, the central device of any network or communication system cannot be expected to command and control the behaviour of the end nodes of a communication network on a large scale and with high efficiency, and the establishment of a large number of "local" connections without the intervention of a central device is an imperative for future networks.[8]

The data of a D2D session is transmitted directly between terminals without being forwarded through the base station, while the associated

[7] CISCO I. Cisco visual networking index: forecast and methodology 2014–2019, white paper[EB/OL]. http://www.cisco.com/c/en/us/solutions/collateral/service-provider/ip-ngn-ip-next-generation-network/white_paper_c11-481360.html.

[8] Qian Zhihong, Wang Xue. A review of D2D technologies for 5G communication networks[J]. Journal of Communication, 2016, 37(7):1–14.

control signalling, such as session establishment, maintenance, radio resource allocation, as well as billing, authentication, identification and mobility management, remains the responsibility of the cellular network.[9]

In the 5G era, the introduction of D2D communication will bring us huge benefits, but it will also face some challenges. When the distance between end-users is not sufficient to maintain proximity communication, or when the conditions for D2D communication are met, the optimal selection of D2D communication mode and cellular communication mode and the switching of communication mode need to be considered. In addition, the optimisation of resource allocation algorithms in D2D communication is also worthy of further study.[10]

M2M Communications

M2M communication is machine-to-machine (M2M) communication. US consultancy FORRESTER forecasts estimate that by 2020 there will be 30 times more communication between things than between people worldwide.[11] There are two main definitions of M2M, broad and narrow. M2M in a broad sense refers to communication between machines, between people and machines and between mobile networks and machines, and covers all technologies that enable communication between people, machines and systems; in a narrow sense, M2M simply refers to machine-to-machine communication.

At present, the most common form of communication in everyday life is still between people and devices, for example, Internet access is communication between people and machines. In the 5G era, machine-to-machine communication is likely to play an important role.

[9] PENG Tao, LU Qianxi, WANG Haiming, et al. Interference Avoidance Mechanisms in the Hybrid Cellular and Device-to-Device Systems[C]//Personal Indoor and Mobile Radio Communications. Tokyo: IEEE, 2009:617–621.

[10] Zhao GF, Chen J, Han YB, Xu C. A review of key technologies for 5G mobile communication networks[J]. Journal of Chongqing University of Posts and Telecommunications (Natural Science Edition), 2015, 27(4):441–452.

[11] SHAO Y L, TZU H L, KAO C Y, et al. Cooperative Access Class Barring for Machine-to-Machine Communications [J]. IEEE Wireless Communication, 2012, 11(1):27–32.

To give an example of home management. The environmental monitoring network in the intelligent home management system monitors the environmental data of the home and sends the data to the cloud. After comparing the data, it is found that the environmental quality in the home is now problematic and the control system then sends a command to the air purifier, fresh air system, etc. for them to work. There will then be little or no need for communication between man and machine.

The development of M2M is now also facing a number of technical difficulties. The large number of machines communicating with each other can cause network overload, which not only affects the quality of communication services for mobile users, but also causes problems such as difficulties for users to access the network. In addition, M2M communications are flooded with small packets of information, resulting in reduced network transmission efficiency, in conditions where charging is not possible, future 5G networks face the challenge of extending the endurance of M2M terminals.

Information Centre Network

Information-centric network (ICN) refers to the trend towards information-centric networks. The idea of ICN was first proposed by Nelson in 1979. As a new type of network architecture, the goal of ICN is to replace the existing IP.

In contrast to the traditional host address-centric TCP/IP network architecture, ICN adopts an information-centric model of network communication, ignoring the role of IP addresses and even using them as a transmission identifier.[12] Information generally includes real-time media streaming, web services, multimedia communication, etc., and the information centre network is the total collection of these pieces of information.

The information centre network is specifically implemented in the following way:

[12] Zhao GF, Chen J, Han YB, Xu C. A review of key technologies for 5G mobile communication networks[J]. Journal of Chongqing University of Posts and Telecommunications (Natural Science Edition), 2015, 27(4):441–452.

In the first step, I publish a video content to the network and when the nodes in the network receive a request related to the content I publish, they know how to respond.

In the second step, a friend of mine learns about the video, and when he is the first to send a content request to the network, the node forwards the request to the content publisher, which is me, and I send the corresponding content to the subscriber, and in the process, the node with the cache caches the content as it passes by.

In the third step, if another subscriber sends a request for the same content, the neighbouring node with the cache will respond directly to the subscriber with the corresponding content.

Mobile Cloud Computing

In the 5G era, there will be up to 50 billion connected Internet services around the world, as demand becomes more and more diverse and people demand more and more computing power and quality of service from smart terminals, especially in terms of computing, which will reach unimaginable levels.

Mobile cloud computing means is the introduction of cloud computing in the mobile Internet. In the past, mobile devices needed to handle a lot of complex calculations and also needed to do a lot of data storage. Mobile cloud computing moves this content to the cloud, which can largely reduce the energy consumption of the device and can also compensate for the lack of storage resources on mobile devices. In addition, storing data in the cloud, i.e. in a series of distributed computers, also reduces the probability of data and application loss.

In the future, the mobile cloud will be used as a service platform to support a wide range of applications such as intelligent transportation and mobile healthcare.

Software-Defined Wireless Networks

An important challenge for wireless networks today is the presence of a large number of heterogeneous networks, such as LTE, WiMax, UMTS, WLAN, etc. This phenomenon will continue for quite some time. The main challenges facing heterogeneous wireless networks are the difficulty of interoperability, the difficulty of resource optimisation, and the waste of wireless resources.[13]

Simply put, software-defined wireless networking is the use of a common model to define and control wireless networks, making network systems simpler.

How is software-defined wireless networking achieved?

First, the control plane acquires and predicts information about the entire network system, such as user attributes, dynamic network requirements and real-time network status. Once this data is available, the control plane then uses this information to optimise and adjust the allocation of resources on the network, a process that simplifies network management and accelerates the pace of business innovation.

Software-defined wireless networks can direct end-users to better networks or multiple heterogeneous networks to serve them simultaneously, simplifying network equipment and providing programmability for devices, making interoperability between heterogeneous networks easier.[14]

Context-Aware Technology

Context-aware technology is a new form of computing. In simple terms, context-aware technology is an information management system that uses sensors and other related technologies to enable terminal devices to sense the current context and analyse contextual information such as location

[13] FERRUS R, SALLENT O, AGUSTI R. Interworking in heterogeneous wireless networks: comprehensive framework and future trends [J].IEEE Wireless Communication, 2010, 17(2):22–31.

[14] Zhao GF, Chen J, Han YB, Xu C. A review of key technologies for 5G mobile communication networks [J]. Journal of Chongqing University of Posts and Telecommunications (Natural Science Edition), 2015, 27(4):441–452.

and user behaviour to actively provide appropriate services to users.[15] It is adaptable, timely and predictive.

Context-aware technology will make the mobile Internet more proactive and intelligent, pushing the information that users want to know most in a timely manner, rather than having them passively initiate information requests. Context-aware technology can respond intelligently to the relevant needs of users within a framework that meets regulatory requirements, i.e. "network adapts to the service".[16]

Edge Computing

Edge computing is the deployment of nodes with caching and computational processing power at the edge of the network, closely connected to mobile devices, sensors and users, to reduce the load on the core network and reduce data transmission latency.[17]

Take drones, for instance. In the past, if a driverless car was driving on the road and suddenly found a cat in front of the car, it was necessary to send the signal to the base station through the network, and then through the switch to the central control centre, after the centre's calculation to reach the conclusion of the brake, and then returned to the base station, the base station finally gave the signal to the car, such a long transmission chain is difficult to achieve the vision of the 5G era, the time delay of only 1 millisecond. After the edge computing method is adopted, the base station can directly send the brake signal to the car, thereby reducing the delay.

Network Slicing

In the 5G era, different application scenarios have different requirements for network functions, system performance, security and user experience,

[15] ANIND K. Understanding and using context [J]. Personal and Ubiquitous Computing, 2001(5):1–34.

[16] 4G AMERICAS.4G Americas Recommendations on 5G Requirements and Solutions, white paper [EB/OL].[2014-10-23]. http://www.4gamericas.org.

[17] Xiang Hongyu, Xiao Yangwen, Zhang Xian, Park Zhuying, Peng Mugen. 5G edge computing and network slicing technology [J]. Telecommunications Science, 2017, 33(6):54–63.

for example, the network performance requirements for intelligent traffic and video viewing are definitely different. If only the same network is used to provide services, the network must be very complex and difficult to meet the functional requirements of certain extreme scenarios, and the network's operation and maintenance will also become very complex and expensive.

For different business scenarios with different needs for network functions, if a proprietary network is deployed for these specific scenarios, and this network contains only the functions required by the application scenario, then the efficiency of the service will be greatly improved, the network performance required by the application scenario can also be guaranteed, and the network operation and maintenance will become simple. This proprietary network is an example of a 5G slice.[18]

Diversification of the network architecture is an important part of the 5G network and 5G network slicing technology is an indispensable way to achieve this diversification. Network slicing technology will be an important means of backward cooperation between operators and OTT companies in the future, and is an indispensable key technology for operators to achieve new profit models.

The Global Landscape of 5G

When it comes to 5G strength, this is a comprehensive system, not just one or two indicators. So, what are the dimensions of 5G strength to look at?

I believe that there are six dimensions that must be examined: (1) Standards-led capability; (2) Chip development and manufacturing capabilities; (3) The ability to develop and deploy system equipment; (4) Development and production capacity of mobile phones; (5) The ability to develop and operate the business; (6) The capacity of the operator.

[18] Xu Yang, Gao Gongying, Wang Lei. An analysis of 5G mobile network slicing technology [J]. Post and Telecommunications Design Technology, 2016(7):19–22.

What are the most powerful countries or economies in the world in the field of 5G? The three core groups are currently the USA, Europe and China, with Korea and Japan also having some influence. In other words, when you look around the world, the 5G market is mostly divided up between these countries.

Who Is Leading the World's 5G Standards?

When it comes to 5G standards, we may all know the so-called coding battle, some of the netizens not familiar with this field simply simplified the 5G standards into a coding battle, so one of the online arguments is that Huawei and Qualcomm for 5G standards, Lenovo fell to Qualcomm, resulting in the 5G standards let Qualcomm have the upper hand. This is a misleading statement, mainly due to a lack of understanding of 5G standards.

The 5G standard is a complex system, including coding, protocols, antennas and many other aspects, so the International Organization for Standardization has several working groups working on the standard. The approach is that one or a few companies will take the lead, write the standard, discuss it, and finally the many standards will come together to form the entire 5G standard.

In such a complete system of 5G standards, several substandards need to be established, and which country and enterprise establishes more projects, naturally holds the leading power in the whole 5G standards. Who can propose a project? It must be a large country or a large enterprise that has the strength to propose it, or an enterprise that has accumulated technology and has the ability to look ahead to 5G.

The companies that have passed the 5G standard project worldwide are: China Mobile 10; Huawei 8; Ericsson 6; Qualcomm 5; NTT DOCOMO Japan 4; Nokia 4; Intel 4; Samsung 2; ZTE 2; France Telecom 1; Deutsche Telekom 1; China Unicom 1; Telefónica 1; Esa 1. China Mobile 10; Huawei 8; Ericsson 6; Qualcomm 5; NTT DOCOMO Japan 4; Nokia 4; Intel 4; Samsung 2; ZTE 2; France Telecom 1; Deutsche Telekom 1;

China Unicom 1; Telefónica 1; Esa 1. By country, 21 in China; 9 in the USA; 14 in Europe; 4 in Japan; and 2 in Korea. The establishment of the 5G standard is divided among these countries or economies, and other countries basically have no say. Among them, who is the most powerful country, or who is the important leader of the 5G standard? Of course, it is China.

Many people may be very puzzled by the results of this standard project, why China Mobile has 10 projects, more than the total number of projects of one country in the United States, is China Mobile's technology really that strong?

China Mobile has played a major role in the development of the world's 5G standards and has largely influenced the development of the world's 5G standards, not for no reason, but for a number of reasons, the two clearest of which are:

(1) China Mobile is the telecom operator with the deepest understanding of TDD.

When it comes to TDD, this is the basic principle for mobile communications to work in both directions. Bi-directional operation means that the data is transmitted simultaneously. When we talk on the phone, we can speak so that the other person can hear us and we can hear them at the same time, which requires two-way work. To achieve two-way work, the world has two major technical ideas, one is FDD, that is, the use of two frequencies to achieve two-way work, more simply expressed, is to use two tubes to transmit information, a tube up to send data, a tube down to receive data. This method is good quality and efficient, but the problem is that it takes up a lot of resources and has to use two paired frequencies. TDD is a frequency, separated by time, which is commonly described as a tube that transmits information, one time up and one time down. It is not as fast as two tubes, but the advantage is that it takes up less resources.

Technically, FDD and TDD have their own advantages and disadvantages, and it is normal to use different technologies in different times. In the 3G era, the world's mainstream technology is FDD, whether it is WCDMA in Europe or CDMA2000 in the US. China has proposed its own 3G standard, which has also been adopted as one of the international standards, and this is TD-SCDMA. This approach was criticised at the time and TD-SCDMA had a difficult initial development, but the Chinese government was determined to support it and gave the task of building the network to China Mobile, the country's most powerful telecommunications operator. Initially, China Mobile did not want to take it on, but because it was a national task, it could not refuse. After a difficult construction process, TD-SCDMA has finally been commercialised.

On the basis of TD-SCDMA, China proposed TD-LTE as a technical standard in the 4G era. At this point, the whole world saw China's determination, from chip makers to equipment manufacturers all started to support TD-LTE and also TD-SCDMA. It can be said that China Mobile has won a great victory in 4G and achieved a leading position in the market, and the TD-LTE standard has been adopted by operators in many countries.

In the process, China Mobile set up a GTI consortium to promote TDD technology and gradually became the leader in TDD technology. At the same time, among global telecom operators, China Mobile has the deepest understanding of TDD technology and has formed a strong technical accumulation.

Man's plans are not as good as God's. When it comes to the development of 5G, because of the large bandwidth requires more spectrum resources, and spectrum resources, especially high-quality is very limited, so many countries abandoned FDD, to more efficient, higher spectrum utilisation TDD technology, today the world's 5G technology is used by TDD technology. For TDD has more than a decade of accumulation, TDD networking, technical characteristics of China Mobile has a deep understanding and say, in 5G technology to play an important role, it is not only normal.

(2) China Mobile is the operator with the largest number of subscribers and the most complex network.

China Mobile is the world's largest telecoms operator with 900 million subscribers, almost as many as the total population of Europe and more than twice the population of the United States, with a complex hierarchy of subscribers and very different user requirements and characteristics.

China Mobile has built one of the best networks in the world in terms of coverage, not only in major cities but also in rural areas, and its network building capabilities are a model for all other operators to follow.

China Mobile runs 2G GSM, 3G TD-SCDMA and 4G TD-LTE networks, which are highly complex and carry not only voice and data but also a large number of IoT services, so it has a deeper understanding of the network than other operators.

This shows that China Mobile's views and requirements for the future of 5G have become a benchmark for 5G development around the world in terms of global 5G standards development.

Several operators, led by the United States, are pushing NSA in the development of 5G, which is a non-independent network, and the core network and basic network is still 4G, and then in key areas such as the CBD, building a little 5G base station, and then claiming it is 5G. But the fact is that there is no way to fully implement 5G services other than to use it as fibre to provide more bandwidth to areas where traffic is concentrated. The Chinese operator group, represented by China Mobile, has proposed a roadmap for SA, which from the very beginning is to build a real 5G network. Although the investment is larger and the network construction is more complicated, the biggest advantage of this network is that all 5G services can be carried out. Many people may not be aware that the NSA will still have to evolve into SA after it is established, which will actually cost more money and make the network more complex.

In the roadmap for 5G development, Chinese operators, represented by China Mobile, are more aggressive and more far-sighted, with higher technical requirements. In addition, China has equipment vendors like Huawei and a large number of mobile phone manufacturers and business developers, so China has passed the most projects in the 5G standard. It can be said that China is at the forefront of global 5G standards, and any of the substandards and technologies for 5G would be very difficult to adopt without China.

It is important to note that 5G standards are not led by one country or one company, but need to be promoted by many companies from different countries. And within this group, Chinese companies are the most numerous and contribute the most, as the world has to admit.

Which Country Is Stronger in 5G Chips?

Today's communication is a system formed by computing, storage and transmission. To do well in 5G, whether it is a base station or a mobile phone, all need chips. There is still a big gap between China's chips and the world's best, and there are many areas where we need to catch up. So, which country is the strongest in 5G chips?

To understand this, we must first understand where the chips are needed for 5G networks. The management system of the core network requires a computing chip, as well as a memory chip, while many devices such as base stations require special management and control chips. At the same time, mobile phones need computing chips, baseband chips and memory chips, and a large number of future 5G terminals will also need sensor chips. This is a huge system, and in this respect, China is still far from the top of the world. The following is a breakdown:

(1) Computational chip: Computational chips are needed in servers, core networks and base stations, which can be understood as CPUs. Intel is the most important supplier to Huawei and the most important supplier to ZTE. Apart from a few server chips for which China has certain products, the vast majority of computing chips are dominated by American companies.
(2) storage chips: whether servers or cloud, need a lot of storage, 5G's high speed, high traffic naturally need storage. With today's smartphones, memory has been significantly expanded from the original 16GB, with 64GB being the basic configuration. Memory chips are still dominated by the United States, South Korea and China and Taiwan. There are also a number of companies in mainland China that are making a push in the storage sector, but it will take some time to dominate the market. It is believed that in the next five years, Chinese companies will be able to do more in this field.
(3) special chips: in addition to computing, storage, these general chips, in 5G communication base stations and related equipment, there will be some special chips, this area is still the United States dominant. In addition to companies such as Intel and Qualcomm, there are a large number of companies that produce a variety of specialised chips.

China is the largest market for these American companies. There are also a number of companies in Europe that produce specialised chips. China has also made great progress in this area, with companies such as Heisi, Zhanrui and ZTE Microelectronics designing and producing special chips. It can be said that each country has its own strengths in this field, unlike computing chips, which are monopolised by US companies.

(4) Smartphone chips: the most important terminal for mobile communications is the smartphone, smartphone chips, not only for computing, but also for specialised processing, such as GPU for image processing, NPU for AI processing, so smartphone chips must be as small as possible, low power consumption. With the acquisition of the smartphone chip, we can say that we have taken the crown jewel of the chip kingdom. In the 4G era, the most representative companies supplying to all enterprises are Qualcomm and MediaTek. With the increasing technical strength of mobile phone manufacturers, Apple, Samsung and Huawei have developed their own flagship chips and no longer use Qualcomm's chips. But in the 5G era, Samsung's 5G phones still use Qualcomm chips, Apple has been in a lawsuit with Qualcomm, the final result may still use Qualcomm chips, only Huawei 5G chips will use their own. MediaTek will also adhere to research and development in the 5G chip, while Spreadtrum through years of technology accumulation coupled with increased national investment, is 5G low-end chip on the force. Overall, the 5G smartphone chip, the United States has the most powerful strength, but Huawei has been in the flagship products to fight, and in the low-end products on the ZhanRui will also make a difference.

(5) sensors: 5G is the era of intelligent Internet, in addition to computing, storage, control chips, sensors are new opportunities in the field of semiconductors. There are already a large number of sensors in smartphones, but 5G smart terminals will have even more sensors and greater capabilities. In this emerging field, many countries have joined the competition, it is difficult to distinguish between the high and the low, in addition to large semiconductor companies such as NXP, there are a large number of small and medium-sized enterprises hope to

make a difference, and Japan's Murata Manufacturing and other companies also have certain advantages.

In summary, in the field of 5G chips, the United States generally occupies a greater advantage, if not a major accident, will continue to dominate in the future for some time, while Europe has seen a certain decline, China is making efforts to seek a breakthrough, the next 5 to 10 years, the current market pattern will be a major change is difficult to judge, but China is gradually becoming stronger, is an unchangeable trend.

Development and Deployment Capability of Communication System Equipment

For 5G to come out of the lab and become a service that consumers can use, it requires a large 5G network, which is made up of a series of products such as core networks, management systems, base stations, antennas and towers. We call this network a communications system, and all 5G networks around the world must be served by such a communications system. The ability to develop and deliver such a communications system is a testament to its strength.

In addition to the development of 5G communication systems, network planning and deployment in different countries, regions, geographies and climates, as well as continuous optimisation, will be required to provide good service. To give a simple example. The planning and deployment of a 5G network is different in Hong Kong, China, where there is little land and a concentrated population, and in Xinjiang, where there is a large population, which requires a good understanding of the network, as well as the ability to simulate computers and a wealth of experience.

The world's first mobile communications were invented by the Americans, and Motorola was the world's first and most powerful communications equipment company, followed by Ericsson, Nokia, Siemens, Alcatel, Lucent and NEC. In China, there was a so-called "seven countries and eight systems", which referred to the many communications companies competing for the market in China.

In the 2G era, China's own telecommunication equipment was nothing, and only a small amount of equipment was available at a later stage, a huge gap. In the 3G era, Chinese companies such as Datang, Huawei and ZTE began to make frequent efforts with TD-SCDMA, while Huawei and ZTE also increased their research and development efforts on WCDMA through technology accumulation, and their products are very competitive and continue to open up new territories in the international market. In the 2G era, European companies established themselves as world leaders by unifying standards and consolidating their efforts. The US lacked integration from 2G to 3G, and infighting was very strong, with the government swaying between different groups, supporting Qualcomm at one time and Intel at another, especially in the 3G era, as the fight for standards was on the downside and WiMax was a total failure, and the US equipment vendors were hit hard.

In the 4G era, Chinese companies have accumulated many years of technology and further increased their R&D efforts, while at the same time providing high service levels and competitive prices, and are gradually becoming the main force in the market. Huawei has become the king of this era of communication systems, participating in the construction of networks in 176 countries and regions around the world, and being welcomed for the quality of its networks and services, thus becoming the most powerful manufacturer of communication systems equipment in the world. The second is Ericsson, which is the most powerful system equipment provider in Europe, however, it is gradually falling behind Huawei in terms of global share. Nokia consolidated all the fallen companies under its umbrella, including Lucent, Siemens, Alcatel and Shanghai Bell, and thus took the third position. ZTE was fourth and Samsung was fifth. It is worth mentioning that Chinese companies such as Datang are also involved in the system equipment market, as well as Japanese companies such as NEC, although they mainly focus on the local market and lack sufficient competitiveness in the global market.

In addition to end-to-end system equipment, there are also a large number of antennas, small base stations, repeaters and other related equipment, in which Chinese manufacturers are the most numerous.

In other words, China's Huawei, ZTE and Xinke (Datang's parent company) are the strongest players in the world of telecommunications systems equipment.

In the 5G era, if we exclude political influences, there is no doubt that Chinese companies will be the dominant players in the field of system equipment. In the 4G era, Huawei and ZTE are making inroads in the global market, what do they rely on? First of all, the technology is strong, the technology is end-to-end delivery capability, an operator to build a network, it is impossible to do their own technology, it needs a system provider from network planning to network optimisation, and even late operation and maintenance support can provide comprehensive services, this ability to test the comprehensive strength, and even need to provide part of the mobile phone, Huawei, ZTE's mobile phone business is developed in this way.

The technical strength of Huawei and ZTE is world class in the 5G era, which is unanimously recognised by major operators. At present, Huawei and ZTE are under pressure from world powers, which to a large extent indicates that these countries are afraid that Huawei and ZTE will develop and may take a leading position in technology that is difficult to surpass.

In addition to technology, Huawei and ZTE's products are also very competitive in terms of price. One of the keys to a good communications network is to work with people to develop a systematic solution to the problem. In terms of management, Chinese companies are very well known for their efficiency. For the same staff and the same salary, an Ericsson employee works only 35 hours a week, while a Huawei employee may work more than 50 hours. A large number of foreign companies are willing to work with Chinese companies for the same product, but with better service and competitive pricing.

The last point is the service support capacity. It is impossible to say that all communication networks are completely problem-free, and if problems arise, it is a test to see if they can be responded to and solved in time. With relatively low labour costs, high efficiency and long working hours, Chinese companies are far more capable of solving the same problems and keeping the network open in a timely manner than other competitors.

Although Huawei and ZTE have recently faced some political interference in the international market, some Western countries would keep

Huawei and ZTE out of their doors. But I believe that in another three to five years, the accusations that Huawei's network has security problems will be dispelled. Now some politicians do not allow the use of Huawei network, when the operators of these countries will realise that they have suffered a loss, or will gradually turn to Huawei and ZTE, after all, these two companies have strong technology, service capabilities and price advantages.

From the above analysis, it can be seen that the first group in the global communication system equipment field is China, the second group is Europe, and South Korea also has a certain market.

Development and Production of Mobile Phones

The end product of 5G is certainly not just a mobile phone, but for a period of time, the mobile phone is still the more important terminal and will largely influence the customer experience and the development of 5G.

At present, there is a tendency for mobile phone research and development to concentrate on a few enterprises. In today's world, the strongest mobile phone R&D and production strength is the United States, China and South Korea, and the world's top three are Samsung of South Korea, Apple of the United States and Huawei of China. Among the top three, Samsung and Apple are facing certain difficulties. Samsung's brand has been greatly affected since the "battery door", and although the global market situation is still good, the market share in China has shrunk to outside the top ten. Apple's new models in 2018 have not performed as well as they should have because of a lack of innovation. Only Huawei has strong explosive power and good momentum, after consolidating the leading position of the Chinese market, in Europe, India, the Middle East, Southeast Asia, South America markets have a very good performance.

By 2020, it is not impossible for Huawei to reach number two in the world, or even take over the number one spot. By then, 5G will be commercially available around the world, and Huawei's advantages are obvious because of its strong integrated capabilities and its own chips, which can support 5G very well. Samsung is likely to be constrained by Qualcomm in terms of chips, and will also face competition from Chinese companies.

Apple is the biggest problem, is currently fighting with Qualcomm lawsuit, 4G era and even abandoned Qualcomm's baseband chip, if 5G or the use of Intel's baseband, for the accumulation in the field of communications still need to work on Intel, can not well support 5G, so that Apple's mobile phone has a better performance, now there is a question mark, if there is not enough stability in the middle of the problem, that the damage to Apple may be very fatal. In the era of 5G, among the top three global mobile phones, Huawei is the only one with the best market performance, the most solid technology accumulation and upward momentum.

The world's top ten mobile phone brands, China occupies seven seats, South Korea in addition to Samsung, only LG squeezed into the global top ten, while China's oppo, vivo, millet is the top three behind the small three, Lenovo, ZTE although the Chinese market performance is poor, but in the global market performance is good. At present, the first companies to launch 5G phones around the world are mainly Chinese companies such as ZTE, Lenovo, oppo, vivo and Xiaomi. It is safe to say that there is no country that can compete with Chinese companies in the smartphone sector in terms of overall capabilities.

As competition for the Chinese market intensifies, more and more companies are turning to the global market, with Huawei, ZTE and Lenovo having long been active in the international market. In recent years, Xiaomi, Oppo, Vivo and One Plus have all made a push in overseas markets, companies such as Transn are occupying half of the African market and most of the Indian market is occupied by Chinese products.

There is only one European mobile phone brand left, Nokia, but it has been developed and produced in China and is now only a brand of the past.

The first group of mobile phone research and development and production is undoubtedly China, the United States of America's Apple and South Korea's Samsung also has a strong strength, but compared to the comprehensive ability, the United States and South Korea are difficult to compete with China. In the 5G era, China will certainly enhance its strengths even further and consolidate its advantages.

Development and Operation of 5G Services and Applications

5G is not only about networks and handsets, it also requires a large number of services and applications, which is an important factor in whether 5G can be widely commercialised. So, which country is the most competitive in this area?

Traditionally, the Internet was essentially a national copy of the United States. Traditional Internet services were first invented and driven by the US, and then other countries learned from the US, or adopted US services directly.

In the mobile Internet era, China is beginning to catch up and even surpass it. Today's Chinese mobile e-commerce, mobile payment, bike-sharing, taxi-hailing and takeaway businesses, although some of them still have some hints of American products in their infancy, have evolved beyond the original products the further they have evolved.

Two typical examples are given. One is WeChat. The same social product cannot be said to be absent in the US, but WeChat has evolved from a social product to a payment service platform, greatly enhancing the user experience and scenario, which is worth learning from its US counterparts. WeChat offers efficiency and convenience to a large number of users, especially the ability of a large number of small programs, which is incomprehensible to traditional social platforms. Compared to WeChat, Facebook is a far cry from it.

The other is Pinduoduo. E-commerce has been around for a long time, but combining social and e-commerce to create a powerful marketing capability through social is a complete innovation that companies in the US or other countries cannot understand. Today, in many fields, Chinese companies have become the object of imitation in the world.

5G is the foundation of the smart Internet, which requires the integration of mobile connectivity, smart sensing, big data and smart learning, and naturally requires the development and production of smart hardware. Today, China has the largest number of companies with the strongest R&D and production capacity for smart hardware products, such as smart bracelets, watches and body fat scales, and is quickly becoming the world's number one country. At present, Xiaomi's ability to integrate products and the number of products it has access to in the smart home sector is far greater than that of Apple and Google.

Huawei has also started to make efforts in this field, and has formed a strong Huawei Smart product series, through the HiLink protocol, to integrate a variety of smart home, it is not to do all the products themselves, but through the platform, the output integration, intelligence, sales and service capabilities.

The development and production of smart home products in China is among the best in the world. Chinese companies' environmental monitoring products can monitor temperature, humidity, noise, PM2.5, PM10, formaldehyde, TVOC, carbon dioxide 8 indicators, and can be linked with air purifiers and fresh air units for intelligent control, and the product price is less than 100 U.S. dollars, such products are difficult to find competitors in the world.

A large number of Chinese companies are gearing up to develop products in specific areas for 5G networks. From chips and modules to smart hardware, from various smart homes to products for public services and social management, Chinese enterprises are highly motivated and invest heavily, and local governments are also very concerned about intelligent products that can improve the efficiency of social management, such as products for water pollution control and environmental monitoring.

China's innovation in the field of mobile Internet, the accumulation of funds and talents, the research and development of intelligent hardware, production capacity, is difficult to be reached by other countries, which is also the basis for the great development of 5G services and applications in China.

Network Deployment Capabilities of Telecom Operators

One of the keys to the development of 5G networks is the ability of telecom operators to deploy their networks, and only with good deployment can the general public have access to them and related services can be developed.

China's three telecoms' operators are among the strongest in the world, with China Mobile having 900 million subscribers, the largest in the world and almost the entire population of Europe combined, and China Telecom and China Unicom also ranking among the world's top telecoms operators in terms of subscribers.

China's telecoms operators have a strong network deployment capability. Today, 99% of China's subscribers are covered by 4G base stations, and the three telecoms' operators have over 3.5 million 4G base stations, with a total of over 6.4 million base stations, a number that is unmatched by any other country. The US has no more than 300,000 4G base stations and India has no more than 700,000 total base stations. In terms of the number of base stations, China has 10 times or more than any other major country. A high number of base stations means good network coverage and good quality of network. People who have been abroad generally feel that in many countries in Europe and the United States, there is no network not far from the city, or the quality of the network is very bad, in the indoor, the signal in many places is also unstable, indicating poor network coverage.

The most commendable aspect of China's telecoms operators is the excellent coverage of 4G networks in remote areas, with 99% of users enjoying the convenience of the network. The network coverage in rural areas not only bridges the digital divide, but also plays an important role in promoting local socio-economic development.

The gap between other major countries and China in terms of network deployment is very clear and this gap will still be present in the deployment of 5G networks, which will ultimately affect a country's ability to manage society and the efficiency of society as a whole.

In addition to the number of base stations, China has opted for a more aggressive SA-standalone solution in its 5G technology roadmap, while most countries in Europe and the US have opted for an NSA solution. One of the main features of the NSA non-independent network solution is that the main network is still 4G for a long time, and only in the core area with 5G network, that is, this network can not realise all the 5G scenarios and services, it is still 4G network, only in a few places through 5G to improve some speed. The SA solution chosen by China's three telecom operators is to establish an independent 5G network from the outset. This network will not only achieve high speeds for Internet access in key areas, but also support low power consumption and low latency, which provides communication capabilities for industrial Internet, intelligent transportation, etc., and also provides opportunities for the explosion of smart homes.

Imagine, in Europe and the United States a little remote places telecommunications signal is very bad, can not support 4G, and even some of the tourist resorts are not mobile communication signal, we can well understand why mobile payment business can not be carried out in Europe and the United States, the fundamental reason is the lack of a high-quality, high-coverage network. For Chinese users, the default is that they will have Internet access in most places and will be able to pay easily with their mobile phones, a scenario that is not really possible in many countries.

Today, the obstacles and suppression that China is encountering in the process of 5G development are to a large extent not only a suppression of Chinese companies such as Huawei and ZTE, but also a reflection of the anxiety of the European and American elites that they will lag behind China in terms of social management capabilities and social efficiency once China's 5G network is fully deployed.

China will definitely build a large-scale and high-quality 5G network as soon as possible, which will further improve the efficiency of Chinese society, continue to enhance social management capabilities, and make social services more convenient. For many other countries, however, these will be difficult to achieve for a longer period of time. This is true not only in the United States but also in Europe, especially in Southern and Eastern Europe, as most telecom operators lack sufficient funding and governments lack the commitment to build 5G networks.

Government Support and Market Capacity

5G is a huge system project, which is obviously difficult to build without government support and with only corporate investment. For example, there is a need for strong government support and assistance in terms of laws and regulations.

The Chinese government is very clear in its attitude towards 5G development, actively supporting the acceleration of 5G construction, which on the one hand can boost social and economic development, and on the other hand can also enhance social efficiency and reduce social costs.

A very typical example is the allocation and planning of spectrum. Spectrum is a fundamental resource necessary for the construction of 5G, in the same way that land is needed to build a house. Many countries in Europe and the United States use the auction method for spectrum, telecom operators need to spend billions or even tens of billions of euros to get the spectrum, 5G has not yet been built, operators have a very high debt, therefore, small operators for the construction of 5G is not high enthusiasm, large operators even if the attitude is positive, but carry a lot of financial pressure. The Chinese government, on the other hand, has adopted a spectrum allocation approach, whereby spectrum is allocated to telecom operators based on their needs and technology after consultation, resulting in very low spectrum usage costs and low pressure on telecom operators.

Network building is also a big issue. Telecom operators to build networks, to enter the building community, the whole process is more complex, not to mention the price, the negotiation time cannot afford to delay. The government played a big role in building the 3G and 4G networks in China. As a first step, the government required base stations to be installed on buildings first, which not only reduced construction costs, but also accelerated deployment. With the support of the government, it is easier to access communities and institutions for network deployment, which greatly reduces costs, especially time costs.

Starting with 3G, the support of the Chinese government has worked well in the mobile communications sector. In particular, in recent years, under the personal promotion of the top management, telecom operators have carried out large-scale speed and tariff reductions, today China has the cheapest communications tariffs of any major country in the world, and cheap tariffs and extensive coverage have brought mobile Internet to both urban and remote mountain areas. The government will continue to play a huge role in the development of the 5G network. Of course, other governments around the world are also promoting the construction of 5G, such as allocating spectrum and issuing licenses, but there is still a big gap in terms of implementation efficiency and actual results compared to China.

The last force affecting 5G is the market. Whether a technology or a product can be developed, it is important that the market is large enough. Only if the market is big enough can we reduce costs and make capital willing to invest. With a population of nearly 1.4 billion people, China has a particularly high level of consumer enthusiasm for new technologies. This differs considerably from European consumers. For example, Chinese users are far more enthusiastic about smartphones and 3G and 4G technology than any other country, and it could even be argued that the spread of 4G in China seems to have been achieved overnight. While some countries think 4G should be for white-collar workers, old ladies in China have set up neighbourhood groups using WeChat to exchange cooking tips, and smartphones cover all groups from big cities to rural areas.

The market demand of nearly 1.4 billion users is also beyond the comprehension of other countries and regions. While operators in other countries or regions are still achieving high revenues and higher profits through high prices, Chinese telecoms operators are achieving high revenues through low prices, relying on a large subscriber base.

5G is a huge system, and whether it is strong enough depends not on a single point, but on the combined strength of multiple forces. In this complete system, China is in a dominant position in all areas, except for chips. China's chips have also broken the gap, and it is believed that a major breakthrough can be achieved in the 5G era. Looking at the global 5G development pattern, Europe is strong in the system, the United States is strong in the chip, and China is strong in the comprehensive strength. It is expected that with the official commercialisation of 5G, it will be China that leads the world.

A New Option for Telecom Operators

In the face of the upcoming 5G era, telecoms operators have to face up to and be very anxious about how they can build on their strengths and find their own opportunities in the new technology and market situation.

Further Separation of Network, Management and Business Layers

In the traditional telephone era, telecommunications network construction, telecommunications network billing management and telecommunications services were a trinity, fully integrated into a single entity. Telecom operators are both network builders, billing and service managers, and providers of telecommunications services. At that time, the service was relatively simple, with only voice communication and SMS. In this system, telecom operators have full control of the right to speak and control the entire ecological chain. They are very familiar with and enjoy such an ecological chain.

In the 3G era, the telecom network not only allows you to make calls and send text messages, but also to access the Internet. The telecoms operator was very anxious about this change, fearing that it would be reduced to a conduit, and for a long time this was its greatest anxiety. The telecoms operators cannot accept the role of only playing a pipeline and still want to have exclusive access to the eco-chain, in addition to management, they also want to do the business themselves.

Since NTT-DOCOMO started i-mode in Japan, the management and business model has been regarded as the representative model for the development of the industry by telecom operators: the network is provided by the telecom operator, the terminal is customised by the telecom operator and the service is also provided by the telecom operator. In this model, telecom operators unify the world, the network, management, business are done, although the mobile phone is customised, some services provided by the service provider, but the operator has the brand, service, channel strength in all aspects, become the core of the industry chain and business integrator.

With the advent of the 4G era and the ability to access the Internet via mobile phones, there will soon be a large number of business developers who will develop their own businesses outside of the control of the telecoms' operators. These developers were flexible and responsive, with a wide and imaginative range of businesses, and the speed with which the best and the worst were eliminated, so that the system of the telecoms operators was quickly broken down and the i-mode model was largely dismantled.

China's telecom operators have always wanted to establish their own i-mode model, hoping to make a difference in the business, for example, once launched Fetion, payment, mobile phone newspaper and other business was very hot, but over time, the market competition intensified, competitors in the business end of the frequent force, the telecom operators soon fatigue.

It is difficult for telecom operators to operate at the business level, largely because of their systems, thinking, management systems and talent structures, and these problems cannot be changed in a day or two. This is not only the case for Chinese telecoms operators, but also for telecoms operators around the world.

For telecoms operators, the model of relying on network management for profitability is well established and works well. This system is completely different from business development, operation, management and promotion, making it more difficult to integrate. It is not possible to use a different system of evaluation for management and it is difficult to give business development staff a fair market rate for hiring. This makes it difficult for telecoms operators to operate the services they develop, even if they are ahead of the curve, let alone continuously improve and enhance them. Although telecom operators have a strong sales platform, it is difficult to achieve the integration and promotion of new services, which is inefficient. In addition, when there is a conflict between traditional and new business in terms of resources and talents, telecom operators usually choose to preserve the traditional business and abandon the new business. At the same time, a large number of new businesses have legal and social responsibility flaws and many conditions that make telecoms operators shy away from taking action.

At the stage of 3G to 4G development, telecom operators basically accepted the reality of only being a pipeline, although in the video, music, payment in a small way, but the scale of business is very small compared to the pipeline, social impact is also small, in the market competitiveness compared to the first-class Internet companies have a very big gap.

With the addition of data traffic, in the 3G and 4G era, the trend of separating the pipeline from the business is becoming more and more evident, and although telecom operators have gained massive revenue growth through the pipeline, the business is increasingly suppressed by other companies. In the 5G era, operators will continue to do the pipeline or want to make a difference in the business, where the future entry point, thought-provoking.

Compared to the separation of the pipeline and service in the 3G and 4G era, the 5G era will see the separation of the pipeline layer, management layer and service layer.

In the 3G and 4G era, the telecom operators provided the pipes and traffic, and the business developers could do whatever they wanted, and the role of management was not obvious. However, in the 5G era, in addition to the pipeline layer and the business layer, the value of the management layer will become increasingly evident.

Previously, the management of telecoms operators was mainly focused on the subscribers and the role of management was not very visible. In the 5G era, however, the network will not be available to all users equally and without distinction. In addition to ordinary individual users, there will be a large number of enterprise users, and different users have their own requirements for network quality, stability, speed, power consumption and latency. At the same time, billing in this system can be complex, and the principles and methods of billing can vary for different users.

In the 5G era, telecom operators must fully recognise the value of management and prepare for the separation of management in terms of technology and management principles through a variety of capacity building, and seek benefits from this separation trend.

Unleashing the Capabilities of 5G Network Management

In the upcoming 5G era, even with the efforts of telecom operators, the business sector is still not its forte. There are several reasons for this: (1) The range of services is so wide that telecom operators cannot bring all of their services to the table. (2) It is still difficult for telecom operators to make global changes in terms of institutions, management and processes. (3) The most critical point is that 5G will penetrate every corner of social life and social management, which telecom operators are not familiar with. Therefore, it is not a good choice for telecom operators to do business, they may start early and end late, and it is difficult to succeed.

It is worth noting that the 5G era requires not only a ubiquitous, high-quality network, but also a more complex, secure management system that can be adapted to different needs.

Unlike 3G and 4G, 5G is intended for a large number of different application scenarios. Previously, all users used the Internet in one scenario, i.e. through a smart terminal for networking, which was either a computer or a mobile phone, with few other devices. In this network, all devices are equal and there is a single management scenario for the telecom operator, which is only billed through traffic. For 5G, the International Telecommunication Union has defined three major scenarios, namely enhanced high speed mobile broadband networks, low power and large connectivity for the Internet of Things, and low latency and high reliability network communications. For different scenarios, the networks provided by telecom operators are different, the quality requirements are different, and the billing modes are different, so the benefits will be different.

Even in the same scenario, the user profile is different and the services required are different. For example, in an enhanced mobile broadband network, the average user only goes online to browse, socialise, trade, watch videos, etc. These services do not have high requirements in terms of network stability and network quality assurance, but are more sensitive to price. For this group of users, telecoms operators need to develop low-priced packages with a relatively low level of service guarantees.

Also in the enhanced mobile broadband network, if the user is doing remote mobile medical services, it requires very stable network bandwidth, and even special requirements in terms of latency, such as data transmission must not be delayed, lagging.

For the two different services above, it would obviously be inappropriate to have the same network speed. This requires targeted assurance of certain services, management of network resources, redefinition of services, increased requirements for security, etc., and the establishment of a strong assurance capability, which is certainly charged at a higher rate, quite different from that of the average user.

In contrast, many of the services in the IoT scenario require low power consumption and do not require high speed, but have very high requirements for security. Telecom operators have used NB-IoT and eMTC to build a new network to support this scenario, and it will naturally have a new system for billing.

In terms of management, the most obvious scenarios are those requiring low latency and high reliability, which are mainly served by intelligent transportation, industrial Internet, unmanned aircraft, etc. It is important to ensure its security through a variety of technologies and to do a good job of securing the underlying layers and the network side to keep all kinds of risks out. At the same time, a variety of techniques are needed to ensure that latency is reduced. When an ordinary user watches TV, some time delay is acceptable, but when a driverless car encounters an emergency, if the braking signal takes 20 milliseconds to be received, the car will continue to move half a metre and there is a risk of a major accident, so this information must be delivered within 1 millisecond. For such scenarios, network adjustments are needed to form a strong guarantee through a variety of capacity building, and the rates are certainly not charged in accordance with general network services.

For a telecoms operator to develop a strong management capability, a complex system needs to be put in place to support it. Firstly, it provides a variety of network capabilities to support different business scenarios. The enhanced high-speed network of eMMB requires the use of various resources, the consolidation of a large amount of spectrum to provide high-speed transmission and also the use of millimetre waves to support communication capabilities, which is typical of 5G technology. But to support the IoT, it is not possible to use the eMMB network, but rather NB-IoT and eMTC technologies, which form different networks. They have different standards and technologies, different rates and support different terminals. This will require telecoms operators to build multi-slice networks, using multiple technologies to form a complex 5G network, rather than relying on a single network to provide generic services, as was previously the case.

In terms of billing, 5G will be more complex than today's 4G. Communications billing has gone through three main billing units: call length, SMS bars and traffic. In the 5G era, however, more billing models will be needed than just these three billing units. For example, the bracelet, which only records human movement and heart rhythm data, has a small amount of data traffic, which is basically negligible compared to the huge 5G high speed traffic, but it takes up code resources and needs to be managed, so it is obviously impossible to be completely unbilled.

The 5G era is the era of Internet of everything, each household may have a dozen or even dozens of devices, if these devices are managed and billed separately, users will definitely find it too complicated, so we need to find a billing model that protects the interests of operators, but also takes into account the interests of consumers, and at the same time does not make consumers feel too cumbersome.

Urgent Need to Rebuild Own Technology Development Capability

Over time, telecoms operators have evolved from high-technology companies to project management, sales and service companies. Technical capabilities are becoming weaker and we are relying on outsourcing for network construction, management systems and business development. The size of the technical R&D staff is very small in proportion to the total size of the company's workforce.

This was understandable and worked well in an era of very mature technology and stable technological development, as costs were significantly reduced and telecoms operators could devote their main efforts to management and sales. However, in the 5G era, telecom operators will not only provide a universal and indistinguishable network, but also provide users with guaranteed network, management and billing services, which will require them to provide service support at all times, understand users' needs at all times, and react at the first time to rebuild their management capabilities, and the previous model of outsourcing technology and then tendering will not be able to adapt to this new change.

In the following two typical cases, we can see that the lack of technology development and the growing distance from product design are plaguing the next step in the development of telecom operators.

Fetion was once a very iconic application for China Mobile, once a major player in the market, and it had the opportunity to grow into a powerful application that would have cornered the market. However, the development and maintenance of FMS is done by our partners and requires annual tenders. There was a time when the technology was relatively stable and FMS had the advantage of being connected to the telecoms network and could send SMS messages, which was welcomed by users and grew at an impressive rate. However, with the advent of 4G and the rapid development of smartphones, users' demand for traffic has increased significantly and the information transmission mechanism of the network has shifted from Internet access to push, which requires new social information systems. At this time, Tencent developed WeChat at a relatively fast pace and developed this product into the most popular social application in the 4G era, and loaded more services on top of this application to become a brand-new service platform. Fetion itself was slow to transform, and with the emergence of WeChat, Fetion's business shrank severely.

The decline of FMS largely reflects an awkward reality: telecom operators do not own the technology themselves, outsourcing is appropriate for stable technology, but in an era of technological transformation, if their own technical capabilities are weak, and do not study the technology and user needs and find a combination of the two, the situation will become bad. Due to the realities of the situation, it was difficult for the partner outsourcing company to invest in and develop the technology during the transformation process. It is only natural that telecoms operators do not have a good understanding of the combination of technology and network, so it is difficult to keep up with the pace of change.

This is particularly the case with the cloud. The telecoms operators own the premises and the network and are actually very close to the users. Because it provides its own network for its customers, it is in a unique position to develop its own cloud business on this basis. However, in practice, telecom operators also face some difficulties in the development of cloud services, because the majority of users need cloud services, not only cloud storage, but also need to establish a management system, which requires targeted optimisation of their own business, and the establishment of a management system for different services. In this respect, the services and technical support provided by telecom operators are extremely poor, so they lose subscribers and are unable to compete with Internet companies.

5G is not a bare-bones network, but requires strong management capabilities to support it. 5G will penetrate into various fields such as social management, social services and traditional manufacturing. When a large number of traditional services use 5G, a new round of technology development will be required to meet the choices and needs of different users. In this process, a great deal of talent, technology and competence is required.

Currently, China's three telecoms' operators already have a technical R&D team of 100,000 people. It is essential to develop research and development capabilities that are closer to the users on the basis of 5G networks, but it is also a fundamental guarantee for telecoms operators to rebuild their capabilities for the new era. However, the establishment of a reasonable talent pool requires reform of the evaluation, management and remuneration mechanisms, as well as a change in mindset. Only by rebuilding their technology development capabilities will telecoms operators be able to remain invincible in the 5G era.

3

5G Will Redefine Traditional Industries

Intelligent Transport

Prior to the 5G era, the development of transport has gone through several stages. The most primitive means of transport were the feet of man, then horses and donkeys, which were domesticated by man, as well as carriages and ox carts, while sedan chairs and animal-powered tools coexisted for a long time, and later, with the advent of the steam engine, cars and trains replaced the primitive means of transport. With the rapid development of civilisation, it has become commonplace for humans to travel to the sky, to the earth and to the sea. In the upcoming 5G era, human transport will become even smarter and more powerful.

Entering the Driverless Era

The movie "Fast and Furious 8" contains a thrilling scene in which cyberterrorists indirectly control thousands of cars by computer, making them rampage through the streets and become weapons of death.

This is not a figment of the director's imagination, in fact remote control of cars is not a new topic. Today, mobile Internet is created by connecting mobile phones via mobile communication networks. In the same

L. Xiang, *The 5G Era*, https://doi.org/10.1007/978-981-99-4534-4_3

way, the ubiquitous car is seen as a terminal, and the car is connected through a mobile communication network to form the Internet of Vehicles.

In brief, the so-called Telematics is the deep integration of various technologies such as information and communication and intelligent vehicles. The key obstacle is that the data transmission speed of the communication system is not fast enough and the transfer of information between things has not yet been truly established. When we enter the 5G era, ultra-high speed transmission will break this bottleneck and autonomous driving will be possible.

At the very least, driverless vehicles rely on the machine brain inside the vehicle to generate a lot of computing with real-time data from the cloud, which can generate up to 100GB of data per hour. This is something that 4G cannot meet, but not 5G, with its high speed (peak speeds of up to 10Gbps), low latency (1 millisecond) and high capacity (equivalent to 1000 times the current capacity), which can truly reduce latency to 1 millisecond and accommodate the bandwidth for huge data processing, enabling easy driverlessness. The aforementioned scene in *Fast and Furious 8* is a fusion of connected car terminals and driverless technology. Through the car network, the final formation of the car and car connection, the data communication between people, cars, roads and clouds, intelligent features are extremely obvious.

It is foreseeable that driverlessness will be a key part of the huge changes that 5G will bring to the transport sector.

If we turn back the clock, we find that for much of the twentieth century, the automobile, encompassing speed, innovation and individuality, was one of the cutting-edge representatives of modern technology. In order to make the car better able to serve people in their daily lives, most countries have created a large and complex system of roads, which has spread to the countryside. Just as Henry Ford changed the face of the industry in 1908 by producing the Model T, innovation is driven by a spirit of discovery that continues to produce better technologies and products. The excitement of driverlessness will once again revolutionise the traditional automotive industry.

In February 2018, California became the first state to allow driverless cars to be tested on the road without the supervision of a human safety driver, meaning that companies like Uber and Google's Waymo are accelerating their efforts to bring driverless cars to market.

On the other side of the world, in China, driverless cars are also making a welcome appearance. According to China News, on 22 March 2018, several uniquely shaped cars were seen moving smoothly on the streets of Beijing, with constantly rotating instruments on top of them, braking, slowing down, turning and driving smoothly, seemingly no different from normal moving cars. However, as you can see through the window, the driver in the driver's seat has his hands in his lap and is driving the car entirely on his own. This is the scene of the first autonomous driving test trials in Beijing. On the same day, the Beijing Municipal Bureau of Communications issued the first temporary number plates for autonomous driving test trials in Beijing to Baidu, and three autonomous vehicles were officially put on the road for testing. In addition to Beijing, Shanghai, Chongqing and many other cities have also issued policies on driverless car road tests, and driverless car projects are springing up, with the industry heating up rapidly.

Whether at the policy and regulatory level or at the technical level, autonomous driving is gradually being improved, and how these vehicles will affect our lives is still a matter of imagination.

As traditional energy sources continue to run out, and as pollution becomes more of a problem, petrol cars on the streets could be replaced by smarter, driverless cars in the future. In this type of car, people can spend more time on work, resting their eyes or enjoying entertainment because their hands and attention are freed up.

For elderly and disabled people with reduced mobility, driverless cars can be a great help and a boon. Today, the convenience of e-commerce has had a profound impact on those who are unable to drive or who find it difficult to move around without external assistance. Imagine if you were unable to get home from a supermarket, shopping centre, health clinic or restaurant due to physical limitations, a driverless car would pick you up and take you home smoothly.

Of course, the most important thing in traffic must be safety. Driverless vehicles are still in the testing phase. On 18 March 2018, an Uber self-driving car struck and killed a woman in Tempe, Arizona, USA, in the first case of a fully automated car killing a pedestrian. One of the reasons for the tragedy was the failure to put the drones in a 5G environment. Because in the 4G era, driverless vehicles will always feel like they are half a beat slower, and perhaps the data has not been transmitted in time for the scene to change, even with 4.5G, the danger is the same. 5G, on the other hand, can greatly improve security due to the low latency of data.

But whatever the case, it is undeniable that the future of driverlessness cannot be underestimated. For example, 20 years ago, people thought a Nokia phone would be perfect, but nowadays there are large smartphones with modern screens everywhere. So the statement that "we are about to enter the age of driverlessness" is not just a statement. By then, the profession of driver may disappear, driving schools may face closure, traffic police may lose their jobs, and gas stations may disappear…

In the 5G era, autonomous driving is set to revolutionise society.

The Road Redefined

In the process of urbanisation, transport is the lifeblood of economic and social development. Nowadays, there is more and more talk about travel and the way we get around has changed dramatically compared to what it used to be. Whether it is the diversity of travel modes, or the convenience, comfort and safety of travel, all have been improved in an all-round way. But a harsh reality is that road congestion, parking difficulties and traffic accidents are becoming more and more serious.

Traffic systems are time-varying, non-linear, discontinuous, unpredictable and uncontrollable. In the past, in the absence of data, urban road traffic was studied in a "utopian" state. But with the development of technologies such as instant communication, the Internet of Things and big data, full data collection and deconstruction of traffic and travel are gradually becoming possible, and a revolution in the transport system has arrived.

It is foreseeable that the development of intelligent transportation synergy will become a trend. Vehicle-Road Collaboration is known as the third revolution in road traffic safety and is one of the important goals of intelligent transportation development. The basis of a vehicle-road collaboration system is the intercommunication between vehicles and between vehicles and roadside equipment in different locations.

With the advent of the 5G era, the connected car will continue to be upgraded. In the early days of Telematics, it only referred to in-vehicle navigation systems with communication devices on board, where vehicles were able to communicate with the backend of the vehicle through the public network to access primary services such as navigation. Existing traffic information systems such as traffic lights, taxis, highways, buses and other systems are independent of each other and do not share data in the background.

The vehicle-road cooperation system mainly uses wireless short-range communication technology to interact with vehicles and roads within one kilometre, so as to obtain micro-environmental information such as the speed and location of surrounding vehicles, thereby judging the surrounding driving environment, predicting the probability of accidents, and realising the functions of police car priority, police assistance and public transport priority to improve driving safety and traffic efficiency.

At the same time, traditional modes of transport are also undergoing change. At this stage, some cars are already capable of semi-autonomous driving, but these cars will inevitably be interfered with and influenced by other non-intelligent cars during the driving process, making it difficult to avoid traffic accidents. At some point in the future, when driverless cars and ordinary cars co-exist, there may be dedicated lanes on some highways or city roads for smart cars, so that smart cars and ordinary cars can operate in an orderly manner, and so that the capacity of the roads can be significantly increased. And when driverless cars have fully replaced ordinary cars, road planning in cities will become even easier—because cars will be able to automatically identify and avoid obstacles.

Another implication of the redefinition of roads is that the roads of the future will be intelligent digital roads, where every square metre of road will be encoded with active and passive RFID to transmit signals that can be read by intelligent traffic control centres and cars, and where underground roads and car parks can be precisely located by means of RFID. With this precise positioning on the road, the intelligent traffic control centre can effectively manage each vehicle and the user can find exactly the car they are looking for. Most importantly, each vehicle will run along a digital track, greatly reducing the likelihood of accidents. Driverless cars only need to spot obstacles in front of them and react in time, without worrying about the impact of lane changes, overtaking or obstacles. The author himself has applied for a patent for the digital management of roads based on RFID.

Imagine a redefined intelligent transport: early in the morning, the smart car comes out of the garage, having been charged, and is instructed by the intelligent traffic control centre to pick up a customer. On the road, the car will follow a digital track that has been planned to reach the guest and take them precisely to their destination. All routes are planned by the intelligent traffic control centre, which ensures high speeds and no traffic congestion because calculations are carried out to determine which vehicles pass where and at what time. The fare is automatically deducted after the guest has alighted from the vehicle. After a day's operation, the smart cars are disinfected and cleaned in the evening at an automatic disinfection station, and then returned to the charging stations in the underground garage for recharging. All this is uncontrolled.

At one time, around 100,000 people were killed in road traffic accidents in China each year, but in recent years the level of management has improved and around 50,000 more people are killed in road traffic accidents each year. As roads are redefined and intelligent transport systems continue to improve, the number of road traffic fatalities will fall to a few thousand, or even a few hundred, each year.

According to the trend of scientific and technological development, the future road traffic system will certainly break the traditional thinking, focusing on the reflection of human sensing ability, vehicle intelligence and automation is the most basic requirements, the casualties caused by traffic accidents is hardly seen, the traffic carrying capacity of the road network will also be greatly increased. The basis for this is, of course, the need to ensure high speed, stability and reliability of the communication technology.

At that time, more advanced information technology, communication technology, control technology, sensor technology, computing technology will be integrated and applied to the maximum, the relationship between people, vehicles and roads will be upgraded to a new stage, the new era of traffic will have real-time, accurate, efficient, safe, energy-saving and other significant features, intelligent transportation systems will certainly set off a great revolution.

Energy for Mass Storage

Historically, the three main foundations on which human society has moved forward have been matter, information and energy. The world is made up of matter, information is the medium of communication, and energy is the driving force behind the movement and change of all matter.

As the main source of energy for human development, the continuous improvement and enhancement of energy storage capacity has not only changed the way people use energy, but has also driven industrial development, technological progress and human civilisation forward.

To give an example of electricity, which is common in our daily lives, because human life follows a certain pattern, the demand for electricity varies greatly during the day and night, and the huge peak-to-valley difference in electricity consumption makes the peak period tight and the valley period excessive. If the electricity is stored in the low season for use in the high season, it will greatly improve the contradiction between electricity supply and demand, and can also effectively alleviate the situation of insufficient electricity in some areas during the summer peak. Solar energy, for example, also requires energy storage systems to ensure the continuous operation of solar energy installations due to the diurnal variation of the sun and the influence of weather and seasons.

In recent years, the energy crisis has been intensifying and the application of new energy sources in the automotive sector is of great significance for the environment in which we live.

In February 2009, a pilot conference on energy saving and new energy vehicles was quietly held, which marked the beginning of China's new energy vehicle industry. Since then, policies and subsidies have given rise to the rise of new energy vehicles. Nine years later, at the 2018 National "Two Sessions", Premier Li Keqiang mentioned the automotive industry several times in his government work report, with "new energy vehicles" being mentioned as many as three times.

In the field of intelligent transport, energy storage is now in the form of batteries. Future energy development depends on breakthroughs in energy storage technology.

In 1970, researchers first invented the lithium battery and since then the energy density of lithium batteries has been increasing and the cost has been decreasing. However, as with Moore's Law, the development of lithium batteries has reached its theoretical limits and researchers are looking for alternative technologies. Battery energy density is now the most important breakthrough for the battery industry, and even for the electric vehicle industry, to take a big step forward.

According to media reports, the Argonne National Laboratory's Joint Center for Energy Storage is developing next-generation battery technology that is five times more powerful than today's batteries at one-fifth the cost.

Energy storage applications in the field of new energy vehicles include the construction of charging piles, V2G (vehicle to grid) and the secondary use of power batteries.

Now, in the policy, technology, market multiple promotion, the development of new energy vehicles accelerated, in 2016 when the mainstream pure electric car range is less than 300 km, now have begun to enter the 500 km mark, and even some models have reached 800 km range.

As countries around the world pay more attention to environmental issues, low-carbon and green mobility are becoming the trend, and there is reason to believe that the future energy mix will definitely be diversified, and traditional energy sources cannot be withdrawn from the market overnight, so multiple energy forms will co-exist for a longer period of time.

Currently, new energy sources, mainly wind, light and water, are discrete and distributed. Increasingly advanced energy storage technologies will gather and store energy produced on a large scale, so that even though energy production methods are diverse and even discrete, energy use will still be centralised and high-density.

New energy vehicles are disrupting the mobility revolution, and there are both opportunities and challenges. For the industry, domestic and foreign technology reserves are strong car companies, battery manufacturers are in the thick of the layout, in order to be able to quickly in this emerging market to stand firm.

There have been many masters in the history of mankind, but in modern times there are basically no more masters, not because of a lack of knowledge, but because today there is no longer a monopoly on information, it is silicon that has changed this. In the days when paper was the storage medium, information was stored in small quantities and disseminated slowly, making it difficult for the average person to access large amounts of information, which gave rise to the learned masters. The emergence of silicon, which has made mass storage of information a reality, has resulted in a time when no one can become a monopolist of information, and so masters are scarce.

The next problem for mankind to solve is the mass storage of energy. Wind, solar, hydro and tidal energy are inexhaustible, but one of the biggest problems is that they cannot be stored on a large scale, resulting in a lot of energy being lost for nothing. Today's lithium polymer batteries are the most powerful batteries available, but they are not dense enough and the power they store takes up a huge amount of space. The fundamental breakthrough in solving the problem of mass storage of energy is new materials. New materials such as graphene make high thermal conductivity, high electrical conductivity and high transparency less of a problem, allowing for a significant increase in charging speed and solving the problem of speed of energy transmission. However, it is still quite difficult to find a new material with high energy density, but also with good stability and safety. If mankind achieves a breakthrough in this field, the changes to the world as a whole will be enormous and could even reshape the global wealth structure and political landscape.

True Car Sharing

The sharing economy is booming, with shared bicycles, shared cars, shared umbrellas and shared rechargeable batteries everywhere in large and medium-sized cities in China. When it comes to the sharing economy, the term was first coined in 1978 by Marcus Felson, a professor of sociology at Texas State University, and Joe L. Spaeth, a professor of sociology at the

University of Illinois, in a published paper. In addition, the sharing economy has a formula proposed by Robin Chase, the originator of the sharing economy: sharing economy = excess capacity + sharing platform + participation by all.

In layman's terms, the sharing economy is the use of resources that are temporarily unused and unused by others, combined with the participation of everyone.

In the field of mobility, in addition to autonomous driving, car sharing is also being talked about. Shared mobility will be another major direction for mobile travel in the future.

According to statistics, most private cars are idle 90% of the time. The rise of YouTubes and Drips has allowed idle private cars to play a huge role in relieving traffic congestion and tackling environmental pollution, and this role will be multiplied when 24-hour autonomous driving and 5G technology are added to the mix.

Traditional taxi hailing and car hire services have changed our relationship with the car to a certain extent, to the extent that you don't have to drive your own car to get to a place. Since its inception in 2010, Uber has expanded to more than 600 cities in 77 countries and territories, providing millions of trips every day and making a clear impact on society and the public.

In 2018, the topic of gunpowder is the successful launch of Meituan, Meituan taxi stationed in Shanghai to seize the market share of Drip, and achieved 700,000 orders in three days; at the same time, Gaode Map announced the launch of the hitchhiking business on March 27 (on 26 August 2018, Gaode Map has taken offline the business). It is easy to see from the moves of DDT, Meituan and Gaode that the unicorns in the field of travel have identified the big piece of cake of shared travel.

In addition, the development of car time sharing has been relatively rapid. This model has been operating in Europe and America for over a decade. In 2013, Shanghai also began promoting time-share vehicles.

It is foreseeable that, in the field of travel, after the official commercialisation of 5G, a variety of services such as private cars, taxis, car rentals and time-sharing will be fully integrated with self-driving technology, providing people with a safer and more efficient way to travel, and the connotation of car sharing will be even richer.

It is foreseeable that the car sharing scenario will be as follows: the user looks for the vehicle at the smart terminal, clicks on I want to get in, enters the location of the car, and then the car automatically pulls out of the parking space and comes to the user's location. Once the user is in the car, the car automatically matches the relevant route, makes intelligent decisions and drives to the destination. When you arrive at your destination, all you have to do is click on "I want to park" and the car will automatically pull into the parking space and wait for the next user.

In the future, people will have to buy cars or not, and private cars will be completely replaced by shared cars, depending on how far these new modes of sharing can meet people's travel needs, and all this can only be tested by time and the market.

It is worth mentioning that the evolution of the automotive industry is taking place, whether it is the gradual replacement of fuel cars by new energy vehicles or the replacement of private cars by shared mobility. The rapid development of intelligent transportation will bring a brand new travel experience for human beings, and "self-driving + car sharing" is the real car sharing.

After the widespread coverage of 5G network, human beings will enter the era of intelligent transportation, and it is very likely that all cars will become shared cars, and only a few people with special needs will be in possession of the cars, and for ordinary people, when they need a car, the intelligent transportation system can send a car to pick you up, and you only need to pay for the service. These vehicles do not require a person to drive them and the cost is very low. These vehicles are congestion-free, efficient, safe and reliable because they are supported by intelligent transport systems.

Therefore, today's taxi applications are, to some extent, providing an opportunity to experiment with future service models, except that these cars are still driven by people today and the costs are still high.

Healthcare

For a long time, people have been more interested in the use of smart devices in healthcare, including remote diagnosis, remote treatment and even remote surgery. This would, of course, go a long way towards addressing the imbalance in healthcare. Another brand-new opportunity in the healthcare sector is intelligent health management, which enables people to achieve physiological coordination, psychological balance, balanced diet, exercise adaptation and environmental cleanliness through intelligent management capabilities. This kind of health management is the integration of big data and artificial intelligence capabilities through intelligent sensing capabilities to manage people's lifestyles and help them to form good lifestyles and habits.

In the long run, the effects of intelligent health management may be more tangible than telemedicine.

Health Information Is Collected in Real Time

In the 5G era, through the Internet of Everything, a person's body will be able to detect any problems in advance, either by himself or by the hospital, and the abnormalities will then be treated and controlled in a targeted manner. In this way, it is possible to largely avoid the situation of being informed of the late stage of XX disease only when going to the hospital for examination.

In fact, before 5G became commercially available, the medical community had already created a number of devices to capture information about human health.

In the beginning it was mainly medical instruments based on Bluetooth and GPRS systems. For example, the patient carries a lightweight wrist blood pressure monitor, which monitors the patient's temperature, blood pressure and heart rate 24 hours a day and transmits the information collected via Bluetooth technology to a mobile phone or a device with wireless communication capabilities, and use GPRS technology, this information is sent to the computer in the monitoring centre, which then analyses the information and takes appropriate countermeasures based on the results. If the data is normal, the computer in the monitoring centre will not process it, but will simply save the information for later use in comparing the information. If there is an abnormality in the data, the computer in the monitoring centre reacts quickly, for example by activating the alarm system.

With the development of the Internet, new types of intelligent medical devices have started to appear, based on technologies that have shifted to the more advanced Internet of Things and cloud computing. For example, real-time monitoring of physical information in the patient's home through terminal equipment such as radio frequency instruments. Through the Internet of Things, hospitals can provide real-time diagnosis and health alerts to patients to effectively reduce and control the occurrence and progression of diseases.

With the advent of the 5G era, the ability to collect health information has become even more convenient and useful. A report by the Haas School of Business states, "The most significant impact of 5G in healthcare is 'medical personalisation'. IoT devices can quickly process, analyse and return information by continuously collecting patient-specific data and recommending the right treatment plan to the patient, which will allow for greater patient autonomy."

With 5G's fast processing power, a wide range of continuous monitoring and sensory processing devices can be better supported, patients can be continuously monitored, continuously informed of medical data and preventative care generally becomes possible.

The intelligent terminal information collection module can collect key parameters of human health, such as airway, respiratory airflow, blood pressure, heartbeat, pulse and other information, and also has the function of intelligent analysis, which can analyse the condition online, and through the wireless network and certain information processing platform, can complete the health information perception of the human body.

We can imagine a person wearing a medical sensor at home, his vital signs can be transmitted in real time, the hospital, the doctor, the person can be the first to grasp, so that the medical plan can be developed dynamically, the human body's medical health is a big step forward than before.

This is happening now. In 2017, a medical IoT product won a prize in the Qualcomm Tricorder XPrize medical device competition, which diagnoses and interprets 13 health conditions. Its data collector is a sensor that can be placed in the hand, allowing people to easily see their health status at home.

Qualcomm Life, Inc. has also landed a 5G-based healthcare IoT platform, 2net, which uses biometric sensors to capture a variety of data about the human body that can be seamlessly transferred to the cloud for integration with other programs or portals for continuous monitoring anywhere, anytime. The 2net has been well received so far, not only is it comfortable and useful, but it is also very affordable.

Of course, this is just the beginning. The 5G era is much easier and broader for the collection of health information. In healthcare, where timeliness and comprehensiveness of information are important, 5G networks can provide the ability to transmit more and more health data in real time. The Internet of Medical Things (IoMT) ecosystem of devices and sensors can help people to personalise their health treatment plans after collecting all the health information available to them. Not only patients, but also healthy people can use the device to monitor their diet and fitness, giving them real-time warnings and adjusting their status when appropriate.

Multidimensional Health Modelling

With the improvement of living standards, people's pursuit of health has taken a new turn. Today, people need not only physical health, but also mental health and spiritual peace of mind. Health management is no longer just about treating a single disease, but has evolved into a process of managing a person's health throughout the life cycle.

The arrival of 5G has made the healthcare IoT ecosystem possible. In the grand scheme of things, this ecosystem could include at least billions of low-energy, low-bit-rate medical health monitoring devices, temporary wearable devices and remote sensors.

By using these devices to provide real-time data on vital signs and physical activity, the doctor or health manager can effectively manage and adjust the patient's treatment plan. This data can also be used for predictive analysis, allowing doctors or health managers to quickly detect health patterns in the general population, thus making diagnosis more accurate.

In the healthcare IoT ecosystem, the biggest changes are coming from hospitals. The traditional treating doctor may become a data specialist and the traditional hospital will become a data centre, which is a revolutionary change for the medical service industry. Hospitals can access big data on patients from a shared database, process it with machine algorithms, and the system automatically analyses and evaluates it to provide sound advice to doctors. Doctors no longer have to study complicated cases to come up with the most sensible treatment plan.

As one of the three main application scenarios for 5G, eMBB will enable an ultra-fast network experience to support personalised medical applications and provide an immersive experience, with virtual reality (VR) and online video being widely used in healthcare. This means that patients with health problems can be diagnosed without having to go to hospital and can be cared for remotely and virtually from the comfort of their own home.

This service is a great way to break down the time and space barriers between doctors and patients, as doctors can simply put on a VR headset or glasses and treat patients remotely via 3D/UHD video telepresence or UHD video streaming, allowing these patients to receive timely care. As experts say, "With advanced monitoring technology, 5G will undoubtedly improve the ability of doctors to stay in touch with their patients, whether outside the ambulance or at the patient's home." Thanks to the low latency of 5G, doctors can even operate on patients thousands of kilometres away using robots.

Here is a real-life example. A hospital on a remote island in France is able to provide remote ultrasound diagnostic services to patients in this remote area through a remote B-ultrasound robot, and connects mainland doctors and clinicians in real time for consultations, thus reducing the cost of access to care. The ultrasound robot, which is already commercially available, is a prime example of force feedback functionality and the use of the "haptic Internet", which allows for more precise remote operation with a signal that requires an end-to-end delay of 10 milliseconds, which is not a problem in a 5G environment.

If a patient is in serious condition, 5G's new radio unified interface ensures that such critical transmissions are superior to other transmissions, allowing the patient to receive timely treatment. For example, a 5G IoMT sensor worn by a heart attack patient could transmit signals and vital signs of his or her distress to a nearby hospital at a higher priority than other transmissions, thus ensuring that he or she receives timely treatment.

IoMT also offers robust security solutions, such as seamless and secure sharing of healthcare data, ensuring that patients' private data is not exposed or exposed to other security risks.

In addition, 5G will lead to the decentralisation of medical resources such as graded care. Specifically, in some remote areas, Internet access will become smoother due to 5G network coverage, and rural hospitals that were previously out of reach of advanced medical facilities will be connected to medical institutions in major cities through 5G to form a network of life, thus being technologically empowered and supported.

It is conceivable that 5G will play a key role in both hospital treatment and home care, and that the resulting multidimensional health model will benefit the majority of people.

Building a Healthier Lifestyle

If we look at the causes of the disease a little further back, we can see that the source of the disease is mostly a poor lifestyle.

According to a survey, one in every 10 people in China is a chronic disease patient, with nearly 200 million people suffering from diabetes alone, and the number is increasing at a rate of 20 million per year. In addition, hypertension, cardiovascular diseases, malignant neoplasms and chronic obstructive pulmonary disease are also commonplace, and the seriousness of these diseases can be seen in the fact that more than 20 million people die each year from them in China.

Embarrassingly, 70% of these chronic diseases are self-inflicted, with poor lifestyles being the main culprit. Statistics show that lifestyle accounts for 82% of the major factors influencing health (including the social and natural environment), genetics for 10% and medical care for 8%, thus demonstrating the importance of lifestyle in the prevention and treatment of chronic diseases.

There is an urgent need for a healthy lifestyle that will lead us to a state of healing before we are sick. Because unlike other consumer goods, health is an "immediate need" for everyone, and as the quality of life improves, so does the concern for health.

A health tracking system in the 5G era could meet this human need.

The health tracking tools used by humans in large numbers are still only available at the application level, for example Ginger.io. It is a tool that provides patients with early warning of illnesses, collects data about the body and then automatically detects if the body is sick and, if so, warns people about what kind of illnesses the body may be suffering from. This app is very useful for older people with autism and dementia, which are not easy to detect and some people don't always tell their family about the disease, but this app can alert you.

There are many more applications like this, but for the time being they are only doing very rudimentary health tracking and not more advanced tracking, i.e. lifestyle tracking.

The connected nature of 5G makes it possible for personal devices to access the Internet at any time. Cloud-based applications can cover all aspects of lifestyle, such as the right amount of exercise, a good diet, the right amount of water … Once the user's lifestyle data is transmitted to the cloud in real time, the cloud can then provide the user with recommendations.

In the 3G or 4G era, this was a very cumbersome process, as the health tracking device had to store the data at the terminal and then transmit the data centrally using a mobile phone or Wi-Fi. As it is more cumbersome to use and the experience is poor, it has become a major challenge to popularise it. In the 5G era, the health sector has a broader imagination with on-the-go devices that can access information anytime, anywhere, low power consumption and direct connection to the cloud.

Perhaps when 5G becomes commercially available on a large scale, devices such as mugs, cookware and watches could be part of building healthy lifestyles, guiding our lifestyles and eliminating the causes of disease anywhere, anytime.

Smart Home

People have long dreamed of making their homes more user-friendly, amiable and humane, of realising the "smart home". In fact, current IoT technology can already connect audio and video, lighting, curtains, air conditioning, security, digital cinema and other devices in your home and provide various control functions. However, it is clear that the current technology is still in its infancy and is far from the ultimate in human requirements.

The concept of the smart home has been around for 20 years, but it has not developed well for a long time, and one of the core reasons for this is that communication capabilities have not been added to the smart home. The intelligent control of these home products is too simple and the control ability is poor, and they cannot really form an intelligent system yet. In this case, most smart home products have a single function and poor experience. With the arrival of the 5G era, NB-IoT and eMTC technologies will be widely used in smart homes, which will usher in huge opportunities.

Environmental Awareness Becomes a Reality

In 1984, when United Technologies Corporation renovated an old building in Hartford, Connecticut, it used a computer system for the first time to monitor and control the building's lighting, lifts, air conditioning and other equipment, and to provide a range of information services such as voice communications, e-mail and intelligence information to the building.

This is the first application of information technology for construction equipment. The building carriers, which were previously outside the information subject and controlled only by manual or verbal communication, seem to have come to life to a certain extent, becoming less inefficient and rigid.

In the world of architecture, this is the beginning of intelligence and the old building has the honour of being called the world's first "intelligent building". Although it is only an old building, it represents the start of a new trend.

Since then, the concept of "smart home" has been widely referred to and defined as "the connection of various information-related communication devices, household appliances and home security devices in the home via Home Bus Technology (HBS) to a home intelligence system for centralised or off-site monitoring, controlling and managing household affairs and maintaining the harmony and coordination of these household facilities with the residential environment."

Smart homes have turned the perception of the home upside down. In the smart home, information interacts everywhere, we do not need to control it artificially, the building itself can do everything for us, people, things and the environment are just a part of this intelligent network.

What the smart home wants to achieve is the automatic capture and adjustment of information, and with the advent of the 5G era, this is becoming easier and easier. In the 3G or 4G era, people's control of the smart home mainly relied on the remote control of mobile phones, while in the 5G era, people pay more attention to the "self-awareness" of smart devices. In other words, the smart home will no longer be passively controlled by the user, but will actively "sense" the environment and react accordingly.

The home environment contains many parameters, such as indoor air humidity, temperature, quality, as well as light and sound intensity. After all, the air quality in cities is becoming increasingly unsatisfactory, and the emergence of haze and dust storms has made people desperate for a home system that can be intelligently regulated. The 5G smart home senses these environmental parameters, analyses them and then automatically links the relevant devices. The major difference from previous smart devices is that they do not require human guidance or remote control—everything is active.

For example, China's QIIZ health technology company has developed the world's most integrated environmental monitor, which can monitor temperature, humidity, noise, formaldehyde, TVOC, PM2.5, PM10, carbon dioxide and other data, and transmit the data to the network through various communication capabilities such as Wi-Fi and NB-IoT. The data can be analysed through the intelligent cloud platform and controlled by the air purifier, fresh air blower, range hood and other equipment in the home, thus realising environmental awareness and intelligent management of air quality.

Huawei's smart home platform, through the HiLink protocol, a variety of smart home products connected, lighting, cleaning, energy saving, environment, security, health, kitchen appliances, audio and video, bathroom and other types of equipment through the HiLink protocol gradually open up, interoperable, forming an intelligent service system.

With the arrival of 5G, the smart home will explode, a large number of devices in this field already have the basis of intelligence, just need a low-power communication capability to join, it will be able to change the industry landscape to a large extent.

More Reliable Safety and Protection

The foundation of the smart home is the Internet of Things (IoT), and one of the major advantages of the IoT is the ability to integrate what we originally thought of as "high class" enterprise applications into the family home, such as security protection systems, which can maximise the elimination of various hidden dangers in home security.

Traditionally, the security system of a smart home is a comprehensive application of sensor technology, radio control technology, fuzzy control technology and many other technologies.

In the 3G and 4G era, most common home security systems take the following form: the user installs a camera in the home and sets up an intelligent control program. On top of this intelligent control, users can monitor their home from anywhere, at any time, via mobile phone or tablet. In the event of an emergency, the user issues the appropriate command and the intelligent terminal in the house receives the command and takes prompt action to control the situation. In addition, window sensors, smart doorbells and smoke sensors are all part of a home security system that has some limits built into the system itself or has a camera that alerts in the event of a dangerous situation and allows the corresponding terminal to be turned off manually or automatically by a person.

These security systems have gone some way to ensuring the safety of homes and improving people's lives at home, but there are also some problems. For example, if sensitive data is stolen, this could lead to a personal privacy breach or a break-in of the smart home. Belkin, a well-known smart home manufacturer, had several of its products hacked due to signature flaws, with child monitors eventually turning into bugging devices for hackers. In addition, the lack of security protection of the traditional network itself also makes some security systems have security loopholes, the original security protection has become precisely the insecure link, which is actually the most worrying problem for users.

But in the age of 5G, the situation will change completely. The ultra-high-speed transmission of 5G greatly facilitates the detection and management of information, so that the "perception" between the various components of the smart home is more accurate and rapid, and the degree of intelligence will be greatly improved.

Smart home manufacturers in the market now focus on connecting their own products, in order to highlight their own highlights, have made some competitors do not have the product, the industry does not have a standard. By taking the lead in setting international standards, 5G will break the trend of each manufacturer setting its own standards. In this way, the entire system of the smart home will be more stable and, more importantly, it will be a closed system in itself, which will be much less likely to be hacked.

As one of the key criteria for testing smart homes, security has been given high priority and manufacturers are sparing no effort to develop more advanced products. Today, some smart homes in Australia have a large number of highly sensitive sensors built into the system that can easily detect and respond to even a small insect flying around outside the home.

The changes that 5G will bring to smart home security when it is fully commercialised can be illustrated in two dimensions: indoor and outdoor.

For example, if the user leaves, the lights in the room will automatically go out; if a child climbs onto a window or balcony, the system can automatically close the window to prevent the child from falling from a height.

Outside, the security system will automatically switch on when the occupants are away from home or asleep, and in case of an intruder, the system will automatically sound an alarm to prevent the intruder from taking the next step, reducing the damage to the family's belongings.

In other words, in the 5G era of smart homes, the system can easily control all security issues. The security system provides a level of protection against possible hazards in the home, based on high-speed information transmission and accurate logical judgement, after a hazard has occurred, the system will automatically confirm the alarm information, the location and status of the hazard, issue the corresponding instructions and force the line to be occupied if necessary. In addition, after the commercialisation of 5G, the resolution of surveillance equipment will reach 8K, enabling users to easily access higher definition images, richer video details, and higher value of video surveillance analysis, which are undoubtedly necessary measures to increase security protection.

These functions include not only the security status but also the status of the equipment itself, with the aim of eliminating all possible hazards and making the home safe and secure.

Intelligent Integration into Daily Life

The former richest man in the world, Bill Gates, has devoted a large part of his book *The Way of the Future* to a mansion he is building on the shores of Lake Washington.

The book was published in 1995, and the mansion was officially completed two years later, in 1997. On the whole, the mansion occupies an area of approximately 6600 square metres and has a strong "Pacific Northwest Coastal House" style, with its mountainous proximity to the water. The overall cost of the mansion is said to be US$97 million, which is beyond the imagination of ordinary people.

The main thing is that this mansion is a true "smart mansion". In Bill Gates' own words, "It's a 'house' built of silicon and software and constantly incorporating cutting-edge, cutting-edge technology." The mansion is fully built according to the concept of smart home, not only does it have a high-speed Internet connection, all doors, windows, lights and appliances can be controlled by computer, but also has a high-performance server as the management background of the system.

We take a look at the amenities that Gates can enjoy in this mansion.

Because the house is equipped with advanced voice and fingerprint technology, he does not need to activate the key to enter the house, the system will recognise him based on his voice and fingerprints. If it's a hot summer day and Gates wants to enjoy the air conditioning as soon as he walks in the door, he can pick up his mobile phone and connect it to the central computer in his home, use the digital buttons to communicate with the central computer, activate the remote control, turn on the air conditioning, even do some simple cooking beforehand, adjust the temperature of the water in the bathtub and so on.

If there are visitors, they receive a pin with a built-in chip when they enter, which contains information about their preferences and all the devices in the house can be interfaced according to the information contained in the pin. When visitors enter the room, the air conditioning is automatically set to their preferred temperature, the speakers play their favourite tunes and the walls are decorated with their favourite paintings, making them feel at home.

The Gates mansion may be large, but the attention to detail is superb. There are no sockets or switches visible on the walls of the building, and the power supply and transmission fibre are hidden underground, acting as an intermediary between the needs of the owner and the computer in the home, allowing Gates to talk to the computer. All the devices in the home are also able to understand Gates' verbal commands. In other words, the home controls in the Gates mansion are based on a typical digital control.

The Gates mansion is a classic example of a "smart home". The biggest names in technology always have a vision and design beyond the ordinary, and the home is no exception. This mansion, completed in 1997, is still undergoing continuous improvement. But there is no denying that it is forward-thinking and enables part of the 5G smart home, which is to make everything in life as smart as possible.

If previously this vision of the smart home was only possible based on big money from the big boys, then in the 5G era, the smart home could become a reality for many ordinary families.

With the rapid revolution of the information age, some of the previous constraints are being rapidly broken down. The 48 kilometres of fibre optic cable in the Gates mansion, for example, can now be easily replaced by a Wi-Fi module.

For the smart home, which requires different devices to be interconnected, 5G's unique characteristics of low latency, fast buffering and low-power connection make it possible for many household devices to be connected to the system, and it also brings a wide range of all-round progress, and the cooperation between vertical segments and the mobile communications industry will become increasingly close, which is of great significance to the "interconnection" needed for the smart home.

What is certain is that the scenario that can be played out in the smart home in the 5G era is the one mentioned above—where everything in life becomes intelligent.

We can boldly imagine: "In the morning, the comforter and bed couch can wake up the occupants in a very natural state; next, the curtains are opened and the lights are adjusted, when entering the bathroom, the water temperature and even the temperature of the toilet seat are automatically adjusted; the air conditioning is turned on before going out and coming home; when coming home, the hot water is already boiling, the water in the bathtub is adjusted to the right temperature, and the meal is already cooked, The news and information you want to view has been automatically pushed to your phone." The smart home is like a personal butler who takes care of its residents all the time.

After the arrival of 5G, NB-IoT and eMTC technologies will be widely adopted, which will provide users with services far beyond what Bill Gates imagined for the smart home 20 years ago, and the service capability will be more powerful. A large number of gimmicky things will be eliminated, and a large number of products and services will emerge in four aspects, such as security, comfort, convenience and energy saving, and will be integrated in several smart home platforms. And most importantly, it is cheap and accessible to the vast majority of ordinary users, not just the wealthy.

It is not only manual labour that is being replaced.

In 2015, New York Times journalist Nick Bilton recounted his unpleasant smart home experience:

Nick bought a smart thermostat called Nest, set the thermostat to 21 degrees Celsius before going to bed one day, and went to sleep soundly. But unfortunately, a software bug drained the batteries in the smart thermostat and prevented it from functioning, immediately reducing the room temperature to 3 degrees Celsius. Nick's wife and children woke up frozen in their sleep and, in Nick's words, "the house turned into a 'freezer'."

This smart thermostat doesn't work as well as it should. Imagine if the thermostat had a "brain" that could replace the human brain, perhaps this would not have been the case. In the end, this thermostat is not really a "smart" product.

Nowadays, the smart home, the common way of operation requires users to control manually, that is, using mobile phones, tablet computers to display information, human brain processing, and then the results of human brain processing by hand to transmit to the mobile phone or computer, home equipment to receive instructions, and then the corresponding regulation.

In fact, smart homes controlled by mobile phones and tablets can only be described as relatively low-level intelligence. What the smart home really wants to achieve is never just the intelligence of a single product, but to realise the intelligent linkage between different products and systematic operation. In other words, the smart home must have wisdom beyond intelligence, a "brain" that can truly replace the "human brain".

This control center can be regarded as the data brain of the smart home. It is interconnected with all the devices in the home, receives all the information of people and things sensed by sensors, and then thinks instead of the human brain, makes autonomous judgments, and issues the most appropriate instructions. After the device receives the instruction, it performs related operations, so that each device is always in the state most needed by the user.

It is not enough for a smart home to be remotely controlled, it must have the ability to sense, and be able to make independent judgments, coordinate and link up, do the right thing at the right time, and achieve organic coordination of people, things and the environment.

The smart home with "brain" is a 4S ecosystem, i.e. Software (soft), Smarthardware (hard), Supercloud (cloud), Service (service) four aspects, can cover all aspects of people's clothing, food, housing and transport, and make predictions in advance.

What we can see now is that the field of intelligence has made significant progress in three main directions, namely machine learning, computer vision and natural language processing, and these three directions can be applied to the smart home in the 5G era, to achieve the interconnection of everything in the home, the central control of the scene, and the interaction between people and home devices will be faster and more accurate.

What smart home can replace is not just manual labour, but it can give people more flexible and effective care according to living habits and actual needs without any operation, making life more convenient and beautiful.

E-commerce and E-payment

In the 4G era, e-commerce and electronic payments are already playing an important role in people's daily lives. In the 5G era, with the advent of more advanced technology, things will be different again, and some applications may even be more intelligent than we can imagine.

Flat Sales System

In recent years, with the emergence of new concepts and technologies such as big data and artificial intelligence, the ways of business marketing have become increasingly diverse. The traditional marketing model originated in the United States in the 1960s when Jerome Macarthy, a famous marketing guru, proposed the 4Ps theory, namely product, price, place and promotion. This theory has been the rule of product marketing for many years and involves multiple layers of distribution: the manufacturer produces the product and sells it to wholesalers, who then distribute it to retailers and finally to consumers. Before the emergence of e-commerce, this marketing model was once a monopoly in the market, although a wide range of goods, but the various aspects of this multi-layered distribution system is actually fragmented: the consumer's knowledge of the product and the way to buy very different, shopping efficiency is not high; and from the point of view of the manufacturer, the product after research and development, testing, ready to market, you need to first through a third-party channel (such as From the manufacturer's point of view, when a product is developed, tested and ready to be launched, it is necessary to first make the product known to the public through third-party channels (e.g. TV commercials, lightbox ads, posters) and then provide a sales system. The manufacturer's sales approach is generally to distribute to major supermarkets and department stores. This method of selling is still used today, but there is an obvious disadvantage to this method: due to the lack of big data feedback, manufacturers don't really know their customer base very well, so they have to keep on spreading out and trying to cover more areas, but they don't always reach them.

The international fast food juggernaut KFC is a case in point. In the early days of KFC's presence in China, the company was once very strong, and its popularity in the country is evidenced by its appearance in the movie "China Partners". However, due to a lack of detailed research on site selection, the location was considered at the expense of land prices, as well as rising operating costs and product pricing, which led to the closure of a number of KFC shops. In 2005, KFC was exposed to the "Sudan Red Incident" and became a national sensation. The company's head office, Yum!, is actively engaged in crisis communication and has taken measures to quell the adverse effects of the incident through advertising, media campaigns and the establishment of a testing centre, however, "Sudan Red" is still a lingering shadow in the minds of the Chinese people. The "Sudan Red" incident, in addition to KFC's inadequate supervision of food ingredients, is largely a supplier problem, providing seasoning suppliers containing Sudan Red ingredients mixed into KFC's supply chain. In the traditional distribution model, the disadvantages of a fragmented hierarchy are exposed, leading to food safety fears. KFC's response has been a single media campaign, with constant television and other forms of advertising, which has been costly in terms of manpower, materials, money and time.

In fact, KFC is not the only one to suffer from the fragmentation of the traditional sales model.

In 2014, the Foxy food safety incident shocked the nation again. As the meat supplier of KFC and McDonald's, Shanghai Fuxi Food Co., Ltd. was exposed for using poor quality expired meat, through a series of underhanded operations, so that the poor-quality ingredients into KFC, McDonald's, Pizza Hut, Texaco and other nine enterprises. The impact of the incident on food safety in China has been severe, and major food service leaders are facing a serious crisis of confidence as a result.

The traditional sales system of multi-layer distribution, without the support of big data, is like a rounded column with layers of divided partitions, leaving many links in a semi-transparent or non-transparent state, relying solely on the moral integrity of business constraints, hidden dangers abound.

Similar problems are reflected in the traditional retail sector, where there are no very serious security issues, but mainly in the placement of goods.

As a giant of traditional retailers, Wal-Mart supermarket chain's goods are placed very carefully: according to the different categories of goods concentrated in a fixed location, the entrance to the store will be placed in the season's best-selling goods, discounted goods, the most popular goods. However, the placement of goods at the entrance of the store requires long-term experience to make a more accurate judgement, not only to understand the consumer habits, but also to consider the weather, season, time of day and other factors. In addition, determining the relevance of products to products is not an easy task for the traditional retail industry. Wal-Mart, based on experience, will be convenient breakfast and torches such as urgent items placed together, sales, but this is the result of a long time to figure out the results, living in the era of big data people can hardly imagine, Wal-Mart from the occupation of the U.S. market to it later figured out this set of goods placement method, it took many years. In this process of figuring out, Wal-Mart actually less a large number of sales, if you can take this placement method earlier, the money will come in early. Wal-Mart's approach is actually an application of data, but in the traditional sense.

In the 4G era, e-commerce is in full bloom and the leaders in the e-commerce industry have adopted a different approach to sales, with online department stores' Amazon becoming a major competitor to Walmart. Unlike Wal-Mart's one-way marketing, a third of Amazon's turnover comes from unsolicited sales to customers, and the secret of its success is the use of big data: every consumer transaction is recorded in full, and based on analysis, it takes only a few hours to find recommendations for breakfast and emergency items, and all this data is stored in full, which is a great advantage over the fragmented distribution of traditional sales data. This is a great advantage over the fragmented distribution of traditional sales data. Not only can Amazon market products in relation to each other, but it can also infer a customer's income level and household size based on their purchase history and delivery, and then recommend products to them in a targeted manner. In July 2015, Amazon surpassed Wal-Mart in terms of market capitalisation, thanks to the effective application of smart IoT and big data.

It is easy to see that the boom in e-commerce has led to the rapid establishment and expansion of logistics networks, allowing users to see and buy products in virtual cyberspace. Through the Internet and logistics, the modern sales model has broken down the traditional fragmentation, the path from the customer's knowledge of the product to the final acquisition of the product has been opened up by e-commerce, and the sales system has begun to flatten out.

Before the arrival of the 5G era, there are still bottlenecks in the development of e-commerce: because of the deployment of base stations can not achieve full coverage, offline penetration is relatively limited, e-commerce can not be done in all places, logistics and distribution can not be fully covered, especially in remote, inaccessible areas.

In the 5G era, the development of base stations and mobile communications will see a major explosion, the network will be high-density, covering more remote places, the logistics network will also break through the current downward penetration of the difficulties encountered, the distribution methods will also become diversified. By the time intelligent transportation is developed, drones are expected to stand out and deliver to remote areas to achieve full distribution coverage. The distribution channels for goods will break through the various restrictions of time and distance, ushering in a situation where goods are visible and available without the need for multiple layers of distribution. This sales model is also no longer the traditional centralised control, the new sales system will be flattened.

The flat sales model is not inherently disruptive to business activities, but it is a breakthrough in the traditional way of selling.

All Products Are Sales Channels

The sales channel is the process from the production of the product to the final purchase by the consumer, these links are connected to the entire product sales channel.

The traditional sales channel includes many links. Once a manufacturer has produced a product, it generally passes through wholesalers, retailers, agents and trading markets before reaching the consumer. In this long and tedious process, there is inevitably a profit to be made at all stages. Although the goods eventually reach the customer, they take longer, are less efficient and cost more. In the traditional sales channel, the product is the product and the sales channel is the channel, both are different concepts.

The market is full of goods with long sales chains and exorbitant prices. Take an example that the vast majority of people will count. A bag of 65 grams of crisps, the market price of 6 yuan, to the market a catty of potatoes 2 yuan, 6 yuan can buy 3 kg of potatoes, even including the cost of processing, but also see the extent of its profiteering. Of course, the profit margin for snacks is really small compared to that of cosmetics. In addition to going through the basic manufacturers, retailers and transporters, imported cosmetics also go through the tedious process of customs and import duties, and when they finally reach the customer, the price has increased many times.

Let's take the example of a coffee shop. The cost of a cup of coffee is only 2 to 3 yuan, but in the hands of customers is as high as 30 to 40 yuan, which produces a variety of service charges: shop rent, utilities, shop staff salaries, coffee beans and other raw materials of various baking costs, transport costs, etc. For the operator, it costs a lot of unnecessary manpower costs and physical shop operating costs; for the consumer, he or she pays more money. The same happens in the training market: a one-to-one English course charges up to 600 yuan per hour, in fact, the students pay not only for the class, but also for the advertising and publicity of the course provider, the research and development of teaching materials, labour costs (teachers' class fees, school management and sales commission), venue fees, etc. The costs incurred before the product reaches the customer are ultimately paid for by the consumer. For operators, it is difficult to reduce pricing because of the various cumbersome links, otherwise they will lose money.

High prices are one of a number of problems associated with traditional channels. It is also inefficient for product manufacturers, as traditional sales channels are costly in terms of media publicity. Even in the service sector, the distribution channels of the products also pose a problem for operators and customers.

However, with the rise of e-commerce, traditional sales models have been greatly impacted, with various online shops and micro-businesses playing a large role. It is undeniable that traditional marketing channels are still very important because behind the network transactions carried out by e-commerce, there is still product development, transportation, manpower and sales. Today, the organic combination of e-commerce and traditional sales channels can significantly reduce costs, improve efficiency and thus lower product prices. Let's take coffee as an example. If the operator sells coffee in the form of an online shop, eliminating the cost of shop rent, shop staff, some of the costs of traditional channels still exist, but as long as there are raw materials and delivery partners, you can complete sales, pricing can also be appropriately reduced. However, there are limitations to this approach, as raw materials must be sourced at a certain scale in order to reduce costs in volume.

The e-commerce marketing model, in addition to reducing costs, has also greatly improved the efficiency of consumer transactions, as consumers can simply browse, place orders and pay on the Internet, saving a lot of time. In addition, intelligent pushing on online shopping platforms can also expand the sales channels of products and open up the channel for the sale of things. If a customer wants to buy a dress, after searching for the product of their choice, they will see a personalised intelligent push: other outfits to match the dress. If you want to buy a complete outfit, even if the rest of the outfit is from a different brand or shop, you don't have to go to another brand's shop to buy the whole outfit with one click.

For example, on the Amazon shopping site, we can see the Kindle e-reader being marketed. Once you click on the product page, you can see the range of products below: different Kindle models, reader protectors, cases and more. Click on any of these items to see more extensions: phone cases, tempered films, mobile phones, laptops and more. For those with cats at home, you may want to buy a bag of cat food. Once you get to the cat food product page, you will also see cat litter, litter boxes, cat litter, cat climbing frames, cat scratching boards and other complementary products. All these extended products, many of which are different brands and suppliers, do not require the traditional sales model to be understood by consumers, and there is no need to visit all the physical shops in person, just one click to place an order.

Therefore, in the sales system of the Internet of Things era, products are no longer isolated, all products are linked to other similar products, and the products themselves are sales channels.

If this continues, there will even be another situation in e-commerce in the 5G era: the supply side will no longer be centralised. The products we need are not necessarily all made by large manufacturers, but many individuals can use their expertise to personalise products, such as translating, writing or even playing games to get through the game, all of which can become service-oriented products, tailor-made for the customer. Eventually our demand will become more and more personalised, and the supply side will become more and more decentralised, with all products being sales channels, as the products themselves already carry production and flow information, and users can directly customise their orders. In addition to this, there will be a shift in the marketing platform for suppliers: from shopping sites to individual consumers.

All Consumers Are Salesmen

Traditional marketing is based on market research and either face-to-face meetings between salespeople and customers, or media advertising and other forms of promotion, with high-quality products and services to form a more stable cooperation. There are also many one-off sales services, such as taxi rides and catering. In the age of e-commerce, customers receive information about products mainly through e-commerce platforms, such as the aforementioned recommendation pages for complementary and extended products. The emergence of e-commerce has triggered a revolution in the way we shop, and online shopping has entered millions of households. However, there are still many issues that need to be addressed in the e-commerce industry.

In the booming e-commerce industry, consumers are browsing for products and placing orders through e-commerce platforms. The pictures and videos of products we see on the web are materials that are separated from the real thing, which is in the warehouse or with the seller, so the user who sees a product online does not actually see it. This is the equivalent of a seller's show and a buyer's show, where the virtual and the real are separated, which is why there are repeated stories of huge differences between sellers' and buyers' shows. From clothing to takeaway food, from movies and books to service-oriented experiences, there is a wide range of products available on e-commerce platforms. Although in the age of the Internet, information sharing and sharing allows us to see the reviews of various buyers, it is still difficult to distinguish between genuine and fake products due to the existence of "water forces".

Whether it is the merchant's active promotion or the platform's intelligent push, there are certain problems. From the difference between a seller's show and a buyer's show, to the large number of counterfeit goods flying around, disrupting the social and economic order.

Ms. Wan, who lives in Chengdu, has long been in an e-commerce platform to buy clothes, has not encountered any problems, so put down her heart and bought a piece of ripe pu in the site's tea promotion special field. Unexpectedly, the tea will be washed two or three times after, the tea soup is still cloudy, Ms. Wan try to drink a cup, the taste is bitter and rough, that night, stomach pain. Ms. Wan was so worried that she immediately went to the hospital and was diagnosed with acute gastritis and food poisoning. Ms. Wan immediately called the website's customer service, outrage over outright claim that tea is fake, customer service, however, insists that all goods on their platform are 100% genuine, it must be Ms. Wan's first time drinking Pu-erh and she has not yet adjusted to the taste, in fact, Ms. Wan has been drinking Pu-erh for many years. Ms. Wan strongly urged the other party to return the goods and claim compensation, but had no choice, the site had to refund, however, Ms. Wan's health loss and medical expenses were not resolved. It is understood that, after the high-profile rise of the site, bringing in celebrities as spokespersons, and all items are guaranteed to be 100% genuine, however, the reputation of the company has continued to deteriorate as fake products have been exposed, including food, home textiles, electrical appliances and cosmetics. Unlike other e-commerce platforms, this site did not have a user review function for each product in the early days, so consumers had to passively accept the advertisements for the products.

In recent years, CCTV has exposed major e-commerce platforms for selling counterfeit and shoddy products, and a number of drug websites have been exposed, with almost none of the major e-commerce platforms spared. In addition, there are also a variety of seafood networks exposed by CCTV for falsifying delivery notes and producing counterfeit goods, the supervision of e-commerce platforms and market management is imminent. In reality, most consumers rely on their experience to identify the authenticity of goods, and most complaints are fruitless after buying counterfeit goods.

The combination of virtual and real e-commerce can sometimes bring losses to consumers and pose hidden dangers to major e-commerce platforms, as well as to individual online merchants selling authentic products. The lack of systematic management and monitoring has led to public distrust of online shopping due to the discrepancy between virtual and real goods, the prevalence of counterfeit goods and the lack of systematic management and monitoring. Some of the businesses selling genuine products have had to close their doors under the double impact of customer distrust and counterfeit goods. At present, e-commerce is still hot, but there is still a long way to go to get the online shopping market into a healthy development track.

In 1997, the third instalment of the American film series "The Kid" was released and was a box office hit. Unlike the first two films, the third film includes many stunts and high technology that audiences have never seen before, and the enemies have changed from a funny duo to a four-man international bandit, as seen in one scene: when the thieves spotted the owner of the chip in a taxi, one of them raised his hand, clenched his fist and moved his knuckles slightly, and a camera hidden in his glove captured the licence plate number in high definition, to the amazement of the audience. In the days before the Internet, this kind of bold creation was a real brainstorm. One of the scenarios is the electronic scanning of visible things directly, which could become a revolutionary direction for the e-commerce industry in the future.

In the 5G era, there will be a high speed, ubiquitous network, mobile connectivity and a smart Internet that connects everything. In such an environment, higher requirements will be put forward for network supervision and payment trading platforms, and smart sensing is expected to enter a mature stage of development under the conditions of smart Internet. With the rapid development of intelligent sensing and the support of the Internet of Things, the virtual and the real will no longer be separated, but will become one, allowing consumers to have a new experience: the goods on the Internet can directly correspond to the people who use them in reality. Anyone carrying a product can be directly scanned by the smart terminal to identify the make, model and supplier of that product. At such a stage of development, people are connected to things, all consumers are salesmen, and the products on consumers are the sales channels. We no longer have to accept one-way sales pitches from e-commerce platforms and traditional models, as consumers are completely interconnected with each other, and each is a salesperson.

When intelligent identification is widely used in real objects and e-commerce is ubiquitous, electronic payments will also enter a new phase of development.

Electronic Payments Penetrate Everywhere

As e-commerce gradually took hold of the market, electronic payments also emerged.

In the early days of electronic payment, it was confined to between banks, processing bank operations through computers, and then gradually developed to banks and other units processing funds settlement through computers. Electronic payment in the 4G era has broken through the limitations of banks, based on the integration of the Internet and payment systems, real-time payments and transfers can be made using mobile terminals. At this stage, payment platforms for e-commerce transactions came into being. Payment types have also changed from online PC payments in the early days to a wide range of fast and efficient payment methods including telephone, mobile terminals, point-of-sale terminals and ATM transactions.

Initially, online payments required a bank card number, password or security code, or even the use of a special payment tool (such as a bank's U-shield). Although you no longer have to go to the bank to transfer money, as the Internet was not yet developed, there were frequent problems with Internet payments, such as network disconnection, poor signal and payment failure. Such situations are usually tolerated by the general public if they do not occur in an emergency. However, if the payment is very urgent, it can easily be delayed. For example, registration for the National Japanese Language Proficiency Test has been very popular over the years. Due to the limited number of places available, every time before the registration channel opens, many candidates are at the computer, and when the time comes, they immediately crowd into the registration system, fill in the registration form and pay the registration fee online. At this time, many candidates have to wait for the next exam if they encounter unexpected circumstances, such as Internet disconnection, poor signal, or errors in filling out the payment form, resulting in failure to pay within the time limit. The same has happened with IELTS exam registration. Although IELTS has a wider choice of venues and test spaces than the Japanese test, but for students who need to get their results before the time limit, still need to get a place in the exam, and the IELTS exam also requires online payment, if you encounter a similar unexpected situation at the time of payment, it is likely that candidates will miss out on a place in this examination, resulting in delayed admissions, a semester of delay for nothing.

In the 4G era, with powerful network base stations, smart terminals and third-party payment platforms, electronic payments in China (cash payments are still popular in Europe and elsewhere) have become very efficient. With the convenience of payment platforms, people can go out without carrying cash and various cards (savings cards, credit cards, smart cards, etc.) and can settle their payments with just a smartphone, so they can eat, drink and enjoy themselves.

In the face of the new situation, some institutions in the new rise of e-commerce business opportunities, fast, accurate, hard, to quickly seize market share.

The best of these is Alipay.

In October 2003, Alipay was officially launched as a third-party trading platform for Taobao. Positioned as "simple, safe and fast", Alipay provides an intermediary platform for buyers: buyers first deposit money into their Alipay accounts, and only after receiving confirmation of goods will Alipay pay the corresponding amount to the seller. This "third guaranteed transaction model" makes the majority of Taobao consumers feel safe, and Alipay has gradually accumulated its own customer base through Taobao.

In the early days, Alipay was not very ambitious, its development team only wanted the platform to be rooted in Taobao, to be an e-wallet for buyers and to act as a transaction hub to build a foundation of trust between the e-commerce platform, buyers and sellers. But a year later, Alibaba's management, which advocates "embracing change", realised that Alipay could take on more of a role and started working with all e-commerce platforms. At the end of 2004, Alipay was spun off and the Alipay website was launched, embarking on a path to become an independent payment platform.

In January 2005, with Jack Ma's speech at the Davos Economic Forum as the beginning, China's e-commerce entered the era of electronic payment. With the initial customer base accumulated by Taobao and the general environment of the booming Internet e-commerce, Alipay took the external markets of online games, air tickets and B2C as the entry point and actively expanded its users, with the number of users exceeding 100 million in three years. In October 2008, Alipay entered the public bill payment market, taking over the people's daily water, electricity and gas payment business, and reached cooperation with comprehensive e-commerce giants like Amazon and Jingdong, as well as the three major domestic online travel websites (Ctrip, Mango.com and Yilong). At this point, Alipay has doubled the number of users in less than a year. After that, Alipay saw great growth: develop new features such as balance transfers, friend payments, etc; the promotion was launched to benefit the users (a certain amount was immediately deducted from the purchase of the product with Alipay), quickly capturing the market; we continue to expand our cooperation, actively cooperating with domestic and international industries in payment, and gradually expanding our service areas to Europe, the Middle East, Southeast Asia and North America; the areas involved are as small as village kiosks and as large as overseas universities, with more than 300 universities, including MIT and Cornell University in the US and the University of Leeds and Manchester University in the UK, working with Alipay.

In the 4G era, mobile payments have demonstrated their power, bringing unprecedented payment experiences to many users. With the support of a strong mobile network, smart terminals are equipped with mobile payment functions, eliminating the need for users to carry cash and card packs when they go out. Today's code payments have brought a major innovation to electronic payments, linking virtual platforms and bank accounts. In the upcoming 5G era, mobile payments will penetrate into every corner as base stations are deployed more intensively and the range expands rapidly. With the ubiquitous network, information on the platform will be disseminated to the maximum, and we will be able to find the corresponding manufacturer even by searching for products on mobile terminals, completely breaking the limits of production and consumption.

In the 5G era, under the coverage of the smart Internet of Things, the new revolution triggered by smart learning is bound to make electronic payments even more powerful, and the current technology of people relying on scanning QR codes will also be subject to a new round of technological impact.

Proximity Will Become the Next Generation of Payment Technology

Many years ago, people could already see automatic induction doors in shopping malls, automatic induction taps, and later on induction waste bins and induction homes. In the 4G era, smart terminals are sweeping in and smart sensor products connected to mobile phones are coming into view, such as smart bracelets and weighing scales. The main difference between these products and traditional automatic sensor doors and taps is that everything is connected: the use of automatic sensor doors and automatic sensor taps does not involve the concept of data and does not allow the public to associate these products with themselves; unlike smart bracelets and scales, each exercise data, calories consumed and the length of each exercise period are recorded in detail and stored in the mobile phone through Bluetooth connection with the mobile phone, allowing users to see their health data at any time, thus associating these products with their own behaviour and thus having an interest in them.

The most intuitive experience of sensory technology is the physical game.

When electronic games have gone through the two-dimensional virtual plane space of the handle and keyboard, to the powerful Internet operation platform, the rise of the physical game, let people brighten up. In 2006, the Japanese company Nintendo released a new generation of game consoles, the Wii, which for the first time introduced the player's body sensing into the TV game console, combining keyboard, handle and body movements, surprising players: in addition to the traditional handle and keyboard, you can also directly use your own body to control the characters in the game, the player's sense of immersion is very strong. In 2011, Microsoft and Sony launched their own physical gaming products, which were designed to give players a better experience, with games focusing on athletics, sports, fitness and, with the support of the Internet, joining other players for interaction, immediately catching everyone's attention. However, the immovability, the high price, the additional costs incurred (late network platform fees, game software purchase fees) and the imperfection of the Chinese version of these games have not made them very popular in China. This was followed by handhelds and smartphones with additional physical devices, transforming physical gaming from fixed to mobile and allowing players to make the most of their time thanks to mobility. Bluetooth and Wi-Fi on smartphones have made it possible for devices to progress and for scenes from the movie *Top Gun* to be realised.

On 15 December 2017, *The Last Jedi*, the eighth instalment of the American film series *Star Wars*, was released in the United States, and the series has made Disney a fortune with an endless stream of spin-offs. On 20 December of the same year, Lenovo Group officially launched the Mirage AR smart headset package in China, which includes a lightsaber controller, an augmented reality headset and a tracking beacon. In addition to the set, players can access the world of gaming with their own smartphone. The entire game is dominated by smart-sensing technology and is a low-cost version of *Top Gun*. Although the product has not been well received so far, we have seen a major advance in smart sensor technology. In the future, in the 5G era, sensor technology will become increasingly sophisticated, replacing the initial use of many products and even disrupting the way we pay electronically in the future.

In March 2017, Alipay again launched new technologies, one of which is sensor-based payment, which is now used to top up the Beijing bus system. Although transport in the big cities is well developed, bus card top-ups have to be done at automatic top-up machines or at manual windows, often resulting in queues and congestion. Alipay's sensor top-up feature allows people to top-up using Alipay without having to queue up, by simply sticking the top-up card on the back of their mobile phone. Once the mobile phone's own sensor function senses the top-up card, it automatically recognises the information on the card and pops up a top-up page on the phone's screen. Once the user clicks on the top-up button, the amount is immediately credited to the top-up card, which is very convenient.

The connection of everything brought about by sensor technology has already begun in the field of mobile top-ups, and the future of human-object connected payment scenarios is no longer a fantasy.

In the film *Charlie's Angels*, three detectives replicate the corneas and fingerprints of two executives at Red Star in order to successfully infiltrate the company's secret room. Two detectives wearing duplicate fingerprints and corneas entered the identification system simultaneously, and the system indicated that the corneas were incorrectly identified. The detective wearing the corneas is calm and unruffled as the camera closes in on him, calmly and carefully adjusting the position of the corneas in his eyes to make a successful identification.

Fingerprint recognition has only been used for personal identification since 2000 and the general public has not experienced this technology, while cornea recognition has not been common until now. Once proximity technology becomes the payment technology of the 5G era, it is entirely possible that parts of our bodies, such as our eyes, will become the vehicle for payment proximity.

Today's mobile payment technology is implemented by scanning QR codes on mobile phones. By scanning the QR code, the QR code is identified and analysed on the technical side, corresponding to the corresponding product, and then the user completes the payment. And the nature of electronic payments is only related to the user's willingness to spend. In the 5G era, with the maturity of intelligent learning technology, the ability to identify products will become more and more accurate, even through intelligent learning for modelling. The prospect of our eyes being connected virtually and being able to pay with a blink of an eye, without having to go through a mobile terminal, is a cool scenario that is likely to occur in the future with a more developed network. But how can smart sensing identify a user's willingness to pay through body parts? How can proximity payments be secured without a terminal code lock? These are the key questions that need to be addressed in the future.

QR Codes Exit the Payment Arena

At present, QR code scanning is the main method of electronic mobile payment, and this method has become very popular, mainly due to the rapid development of mobile communications. As early as the 1990s, two-dimensional code payment has been popular in countries such as Japan and Korea. As the development of mobile communication technology in China has always been later than that of developed countries, it was not until O2O was fully promoted in China, coupled with the popularity of smart terminals, that QR codes gained a firm foothold in China.

QR code is a way to compile merchant's product price, account number and other information into a black and white graphic record data symbol, which is scanned and identified by an intelligent terminal to complete the payment. This payment technology is very mature, has a large storage capacity, is quick and easy to use, and is relatively low cost, and is quickly gaining popularity. But the payment technology also inevitably raises security issues.

According to the Global Mobile Security Report for the first half of 2012, QR codes have become a new channel for many mobile phone viruses to spread and phishing websites to profit. When users use QR code scanning, they sometimes see their mobile phone swipe out the link address, and the address is bundled with software downloads, which often carry viruses and enter the user's mobile phone if they are not careful. There are even viruses disguised as fee-sucking Trojan horses that use the QR code channel to spread. Once downloaded by the user, the Trojan enters the user's mobile phone and siphons off a large amount of phone bills. For this reason, security has always been the number one issue in network management.

As well as spreading malicious scams through the QR code channel, some unscrupulous individuals are also taking advantage of consumers' greed to devise more elaborate scams: after a certain number of likes are collected, the user is asked to give a screenshot of the likes and pay for the postage. After the user pays for the postage, the criminals ask for a screenshot of the payment code, and once the user gives the screenshot, the money in the user's WeChat electronic account will be looted. The fraudsters take advantage of the fact that many users do not quite understand the function of payment codes (payment codes are not only used for direct scanning by merchants, but also equivalent to bank card numbers + passwords), using the screenshots of users' payment codes to spend heavily, coupled with the disadvantage of scanning payment codes without re-entering payment passwords (within $1000), so that users who are trapped suffer losses.

Mobile payments in the 5G era will usher in induction technology, which will place higher demands on the security of the payment environment in an environment where the security system is being reconfigured. Current security vulnerabilities will have to be fixed in the 5G era, and proximity payment technology will continue to improve. When we can use our body parts to make inductive payments, QR codes and payment codes, as payment intermediaries, may gradually be retired from the history of electronic payments.

Industry and Logistics Transformation

Every change is a revolution in thinking and a technological innovation. To date, human society has experienced three industrial revolutions, the first of which had the most profound impact.

Many people in the country read an article called "The Story of Watt and the Teapot" in their schoolbooks as children: in a small town in England, every household has a fire to boil water for cooking. A young boy named Watt watched the kettle boiling on the stove while his grandmother cooked, and when the water boiled, the lid was lifted and the water bounced as it boiled. Curious, Watt asked his grandmother why the lid was moving and she said that it was because the water was boiling. However, the grandmother was unable to answer Watt's further questions, and ignored them. Curious and inquisitive, Watt's interest in water vapour grew and led to a revolutionary invention: the universal steam engine.

The advent of the steam engine was the hallmark of the first industrial revolution. In the political context of the establishment of the bourgeoisie, British capitalism developed rapidly, expanding production and absorbing advanced production techniques through overseas trade and colonial rule. In 1765, Hargreaves invented the "Jenny Spinning Machine" and machine production began in England; in 1785, Watt improved the steam engine and promoted the production of machines, bringing Britain into the "Age of Steam"; by 1840, the industrial revolution had been completed with the replacement of traditional manual labour by large machine production in Britain. At the end of the eighteenth century, the industrial revolution began in France and the United States. In the mid-nineteenth century, the rest of the world followed the path of the industrial revolution.

The industrial revolution changed the way people thought from the traditional manual to the mechanical, which in turn led to more inventions and changes: at the beginning of the nineteenth century, the Englishman Stevenson invented the train; in 1843, the British inventor Charles Thurber (1803–1886) created the rotary typewriter. Mechanical thinking has led to industrial innovation in all sectors and has become the mindset of problem solvers: Swiss watchmakers produce exquisite mechanical watches; the Germans used mechanical engineering to create the Z1, a programmable computer, and even an Ulysse Nardin watch that played music.

The First Industrial Revolution revolutionised the textile, coal mining, metallurgical and mechanical industries and saw the rapid expansion of once unknown towns into major industrial cities, such as Manchester and Birmingham in England. At the forefront of technological change, the

UK benefited from this and became a world leader, maintaining the myth of the "sunset".

The first industrial revolution opened the door to a technological revolution, followed by the second industrial revolution, which saw human society move from the steam age to the electrical age. The use of the internal combustion engine and electricity became the dominant forces of the second industrial revolution, leading to the development of transport: cars, aircraft and ships were strongly established.

The third industrial revolution, which began in the 1950s, was dominated by "computing" and information technology. Cutting-edge technologies such as computer technology, nanotechnology, aerospace technology, nuclear technology and genetic technology are leading the competition for comprehensive national power between major powers. The American film "Hidden Figures", released on 6 January 2017, tells the story of the struggle between the United States and the Soviet Union in the field of aerospace, and features a large number of scenes of calculation and data verification, which bring to life the important role of "calculation" in the struggle between the great powers. The development of electronics and information technology has brought the "Silicon Valley" in the United States and Zhongguancun in China into the public eye.

The current human society is at the end of the third industrial revolution, while the upcoming 5G era may open up a fourth wave of industrial or technological revolution. Unlike the two previous technological revolutions, the third industrial revolution saw the emergence of cross-disciplines that combined industry and technology to bring a new technological revolution to mankind. In the future 5G era, industry and logistics will be disrupted by a new technological revolution.

In July 2010, the German government released the "Germany 2020 High-Tech Strategy", formally introducing the concept of "Industry 4.0" for the first time. The German Federal Ministry of Education and Research and the Federal Ministry of Economics and Technology are jointly funding an estimated €200 million to support the smartening of industry and manufacturing. In 2014, Chinese Premier Li Keqiang visited Germany and the two sides published the "Sino-German Cooperation Action Plan: Shaping Innovation Together", which officially announced the Sino-German cooperation on "Industry 4.0".

"Industry 4.0", the fourth industrial revolution, the leading force of the revolution is intelligence, after the steam age, the electrical age and the computer and information age, human society will move towards the age of intelligence in the fourth industrial revolution. The three main themes of the smart era are smart factories, smart production and smart logistics.

The Unmanned Workshop Becomes the Basic Production Model

KUKA Robotics GmbH was founded in 1995 and has since grown rapidly to become one of the world's leading manufacturers of industrial robots, with products sold in North and South America, Asia and Europe. KUKA robots are mainly used on the shop floor in various areas of industry and manufacturing to carry logistics and transport, replacing a large amount of manual labour and the human-machine production model. In today's advanced Internet environment, KUKA robots are interconnecting machines on the shop floor of several companies, greatly increasing productivity.

In the face of the advent of the robot era, coupled with the intelligent era of big data and other latest technologies, the domestic home appliance industry has begun to turn its attention to the future, Haier, Midea, Gree three major home appliance companies are actively seeking the transformation of intelligent home appliances.

In January 2017, China's Midea Group acquired a 94.55% stake in KUKA Robotics. Before that, Midea had already set up a fully intelligent production line in Guangzhou, and the workshop production of its air conditioning products had already introduced robots for standardised operations: all workpieces have their own barcode, and under the Internet of Things, through barcode identification, the robot can receive the corresponding identification information to complete the assembly of the corresponding parts and models, but also through information bundling, data analysis, timely detection of problems and errors, management efficiency greatly improved. The road to discovery is always full of bumps, and the introduction of robotic production is a bold move by Midea, as the company needs to spend a lot of money to invest in its operations, and the transformation of traditional home appliances into intelligent ones is also costly. Throughout the home appliance industry, Midea Group has taken the lead in vigorously implementing robot workshops. In the future 5G era, in the trend of Internet of everything, according to the current development momentum, the United States Group is likely to break out of the three major home appliance companies each for the situation of the king, in the home appliance market alone.

According to "Science and Technology Daily" reported on 10 April 2014, the United Nations Economic Commission for Europe and the International Federation of Robotics statistics show that from the second half of the twentieth century, the world robotics industry has maintained a steady growth of good momentum, the world industrial robot market prospects are very good: between 1960 and 2006, more than 1.75 million industrial robots were installed worldwide; since 2005, more than 100,000 new industrial robots have been installed worldwide each year; since 2008, the global installed base of industrial robots has exceeded one million units, with approximately 1,035,700 units, and this figure is still growing.

The International Federation of Robotics predicts that by 2020, global industrial robot ownership will grow from 1.288 million units at the end of 2016 to 3.053 million units.

The use of industrial robots has led to the emergence of unmanned workshop production scenarios, and with the efficient operation of intelligent Internet, production efficiency will reach unprecedented heights.

In China, Foxconn is also a strong advocate of the unmanned workshop production model.

On 17 November 2016, at the 3rd World Internet Conference, Foxconn's President Guo Taiming gave a presentation entitled "Smart Manufacturing Leading the Development of the Digital Economy", which stated that Foxconn now has several factories that can turn off the lights for production with the support of big data from the Internet of Things.

In the future, when the smart Internet is in full swing, the unmanned workshop will no longer be a pilot factory for a few companies, but a regular production model for most manufacturers. Robots used in industrial and manufacturing production will become increasingly sophisticated and smaller in size, leading to a rapid increase in production efficiency and much higher quality. At that point, the human-machine model will no longer be necessary and labourers will be completely replaced by robots.

Customised Production on the Rise

In the data age of advanced e-commerce and intelligent production, consumers are experiencing and accepting a series of new changes, from the early surprise of many new business models to the later habitual, consumers' tastes are upgrading and consumer needs are beginning to diversify. In the past, manufacturers have mass-produced goods for the market and have exported a single product to consumers, and if these products do not contain the colours and styles that customers expect, two things happen: one is that consumers scrape together a product that slightly meets their expectations; the other is that they give up on the purchase. This is a common occurrence and there is a feeling of helplessness for both manufacturers and users. If the user buys the product from home, he or she will not be able to use it for long because the product is not to his or her liking. In addition, if a large number of products are not purchased, they will become stock, resulting in a waste of resources, which is undoubtedly a great loss for the manufacturer.

In the 4G era, the popularity of e-commerce platforms, the full blossom of online shops, and the trendy personalised labels have brought diverse consumer choices to the public. With the rapid development of the economy, the younger generation advocates individuality and they are in urgent need of products that show their personality, so the needs of consumers are becoming more and more personalised. The emergence of e-commerce can meet such individual needs, sellers through communication with buyers to understand the buyer's demand for products, you can customise the production, and such production is completely free of mass production.

In addition to direct communication with the shop, a wide range of matching menus are available on the online pages for customers to choose from when purchasing products. For example, if we buy a cup of coffee on a takeaway app, we will see that in addition to a choice of flavours, we can also choose the proportion of sugar and milk to suit different consumer tastes; if you don't have the option, you can also specify your taste requirements in the comments field set by the shop. This personalised product customisation model has thus allowed the takeaway market to explode.

With the support of big data and the Internet of Things, some goods are gradually being transformed from mass production models to personalisation models. Manufacturers use the Internet to open customised platforms for users, using big data to classify and count customer demand, the cost and quantity of raw materials, inventory management and cash flow control, to deploy sales strategies based on the data, and to customise production solutions according to the individual needs of different customers. Many top tier brands have also joined the ranks of customised production, including Nike and Adidas who have launched personalised, private sneakers; Burberry has teamed up with DreamWorks Animation to launch a bespoke scarf through visualisation technology; The company has launched customised clothing and wear services with many of the world's leading brands.

In 2016, the American luxury brand Kate Spade acquired the bespoke brand Bag Bar. Since 2017, Kate Spade has launched the Bag Bar, a bespoke platform for customised handbags. The Bag Bar is renowned for its customised production, with the possibility of replacing covers and accessories to suit the customer's individual needs. With the acquisition of Bag Bar, Kate Spade will focus on its own handbag business, using the Bag Bar customisation system to cater to the individual needs of its customers. When customising a product, customers can choose not only the raw materials used to make it, but also the edges, gaskets, decorations and so on, with different options at different price points, allowing them to have their own handbag without having to do it themselves.

In the 5G era, small-scale customised production models will not only gradually replace traditional mass production, but will also usher in new industrial changes in logistics technology. In the future, with the powerful support of the intelligent Internet of Things, big data will be fully utilised, all goods will be equipped with sensors, and logistics and distribution will be transparent. Instead of checking whether an order has been shipped and when it will arrive at the next site, buyers can directly track in real time exactly where their goods have arrived and at what speed they are moving in which direction. Previously exposed as a forgery of international courier bills and tariff declarations, in the future, under the mechanism of open and transparent logistics, will be eliminated.

In addition, the addition of big data will eliminate security gaps in the future management mechanisms of industry, manufacturing and logistics distribution.

Every Step of the Way Is Managed

At the Hannover Messe 2018, exhibitors, media and people from all over the world saw high-tech scenes of human-machine cooperation and machine-machine interoperability made possible by information and automation technology. As the host country and the first country to introduce the "Industry 4.0" concept, Germany showed its strong research and development strength to the world. Digital intelligent production, networked energy systems and intelligent logistics solutions were the highlights of the exhibition, showing the future trend of integrating industry, manufacturing and logistics into the information industry.

As a major manufacturing country, China mainly exports textile products, apparel, footwear, mechanical and electrical products, etc. "Made in China" has long been the main way for China to enter the world market. However, a harsh reality is that, so far, most of the "Made in China" products are low value-added products, and if they are joint ventures with foreign investors, the foreigners control the core technology and sales channels, and the domestic production line earns very little profit, and the Chinese side has to pay expensive royalties. With the rapid development of the economy and the development of e-commerce, the traditional manufacturing industry is facing difficulties such as overcapacity, low added value, heavy tax burden, lack of high-end talents and high operating costs. In terms of management, the traditional chain involves a process of raw material procurement, production, quality control and marketing. Such a chain not only has a long cycle time, but is also very redundant in the upstream and downstream, supply side, storage, dispatching and other links, and due to the fragmentation of the links, information transfer is inefficient, and if problems arise, it may lead to unnecessary losses.

When the e-commerce platform combined with the rise of a new model of private customisation, the past product-oriented one-way output of low-end manufacturing began to reveal the drawbacks, some small enterprises more and more difficult days.

In 2017, an article entitled "Manufacturing is dead, and discussing the future rise of China" hit the Internet, in which the author used his uncle's manufacturing plant as an example to describe the tragic state of low-end manufacturing in China, which is in a state of collapse in the data age. The factory mentioned in the article is mainly engaged in the processing of headphones, and has established long-term cooperation with Taiwan customers, but due to the late increase in labour costs, the profit margin is rapidly shrinking, and the factory owner also seeks to transform, the equipment will be automated to upgrade, but cannot meet the requirements of customers. However, when the boss had no choice but to adopt the traditional labour-intensive management model again, he searched for an agent when launching the products, but when he met someone, the agent frequently defaulted on payment or fled with the money, coupled with the cruel competition in the market, a large number of products were backlogged. In the end, the factory was forced out of business when the Taiwanese customer abandoned the partnership and shifted its focus to Vietnam where labour costs were lower.

The story is a poignant one, in which the owner of the factory is trying to keep the factory running, looking for ways to do so, and trying to run it with responsibility for the business and for the employees, but the results are not as good as they should be. If manufacturers can change their thinking, look beyond the traditional mechanical mindset and make full use of big data technology, the outcome could be very different.

The management of traditional industrial production has long been influenced by the mechanical mode of thinking, which lacks flexibility and variability, resulting in a very "rigid" production management. "flexible" production. "Flexibility" will be the core competence of industrial processing and manufacturing production in the future. Unlike the traditional screw-type production line, "flexible" manufacturing is committed to the flexibility, adjustability and variability of processing and manufacturing, with the ultimate goal of maximising production efficiency, optimising the allocation of resources, minimising costs and increasing profits at all stages. The need to transform traditional production has led to an urgent demand for personnel, especially management professionals, in various countries. As a result, many leading universities have established new cross-cutting disciplines, such as "industrial engineering", to train people in this field. The major has become a big hit at major universities, especially at the top US institutions, and requires students to have a high level of mathematical statistics, probability theory, linear algebra, big data analysis, and knowledge of business management. Once such talents leave school and work in industrial and manufacturing processing management, they will be able to connect with the big data thinking management model of the smart era and become the backbone of the transformation of traditional industries in the 5G era.

"Industry 4.0" is a product of the future intelligent Internet of Things era and is an inevitable trend in the development of human society. Supported by the power of big data and the Internet of Things, sensing will become the next generation of technology. The entire chain of industrial processing and manufacturing production will become increasingly refined. The upstream, downstream and supply chain coexist within one information system, and each link is flattened so that any information is known to staff in other links and all management links are transparent. In addition, in the big data storage records, the best managers will use the best data modelling, the best data accounting to maximise profits, all-round efficient management of every corner, the operation of each link to minimise the cost, the real implementation of the flexible production.

At the Hannover Messe 2018, Siemens presented the Dulux digital paint factory, an industrial production unit using digital technology for virtual reality; Bosch presents an intelligent factory operation with artificially intelligent robots for a wide range of complex tasks; SAP exhibited an automated warehouse management system. In Germany, the concept of "Industry 4.0" has already penetrated local companies and many factories have adopted networked machines. The smart manufacturing management model generates more than 10 billion euros of economic growth for Germany every year.

The fourth industrial revolution is already underway and China is taking a proactive approach by strengthening its cooperation with Germany in the development of artificial intelligence. We have reason to redefine the "Made in China" label for the future and believe that Chinese manufacturers will be able to demonstrate their ability to innovate, to set sail again in the era of intelligent Internet of Things, to break the long-entrenched mechanical mindset, to innovate data-based flexible management models and to achieve optimal resource allocation.

More Efficient Resource Allocation

The goal of "Industry 4.0" is to achieve customised intelligent production, so that every consumer can dictate the production of goods according to his or her own will, without even having to operate them personally. For example, the smart fridge at home can sense the decrease of ingredients inside, and based on the purchase information recorded in the database, accurately determine the user's eating habits and tastes, and automatically send ordering information to e-commerce platforms and product vendors, and the shop will automatically receive the order and deliver it to your home in a timely manner.

In the 5G era, if we want to realise the above scenarios, industrial processing, manufacturing production, logistics and distribution must optimise their resources in order to achieve efficient resource allocation.

China's manufacturing industry is facing various dilemmas at present, and the road to transformation seems to be extremely difficult. In the future intelligent Internet of Things (IoT) era of big data, abandoning the traditional mechanical thinking and actively converting to big data thinking, re-integrating resource allocation with a new way of thinking, is one of the feasible ways out for China's manufacturing industry. Currently, many companies and industrial estates are already exploring this new path, with the support of the government, and with great success.

In the second half of 2013, Wujiang District, Suzhou City, first attempted to establish a big data system in all industrial enterprises in the district, trying to accurately grasp the operation of the district's industry through big data analysis, analysis of enterprise efficiency, and the formulation of resource allocation policies. In order to solve the problem of uniqueness of enterprise identification, the Wujiang District Government vigorously promoted the special identification code of the taxation department, the special enterprise registration number of the industry and commerce department, and the special identification code of the power supply company, so as to screen and regulate the organisation code as the unique code of each enterprise, and match it with the codes of other departments to realise the data integration of various units and departments. After several years of efforts, Wujiang District has included more than 16,000 enterprises in the big data system, achieving full coverage of each enterprise. By analysing and evaluating the detailed data of the entire industrial enterprises, the relevant departments have a solid scientific basis for optimising resources, developing new industries and eliminating low-end production capacity, which greatly improves the efficiency of industrial reform promotion. This big data system is called the "Industrial Enterprise Resource Intensive Utilization Information System", through the system, the operation of industrial enterprises in Wujiang, land, energy, output and emissions and other data are clear at a glance. In addition, the system also allows for thematic analysis of data by region, industry classification and enterprise classification.

At present, the Wujiang District Government is continuing its efforts to seek greater policy support and actively implement policies on differential land use tax, industrial land supply, electricity, water prices and sewage discharge, as well as related pilot policies.

In addition to local government trials, some traditional manufacturing industries are also actively using big data thinking to re-integrate resources.

In 2016, Shenyang Machine Tool Factory entered the world's top 500, as a representative of China's machinery manufacturing industry, Shenyang Machine Tool Factory actively seeks to transform, the world's first to develop the i5M8 series of platform-based intelligent machine tools. Unlike traditional rigid manufacturing models, this intelligent machine tool is inherently flexible and computer-controlled, an intelligent system that combines industry, information, Internet, intelligence and integration (this is also the origin of the name i5, named after the initial i of these five English words). This set of intelligent machine tools uses the Internet as a platform for intelligent calibration, intelligent diagnosis, intelligent management, and has the powerful function of intelligent learning, and is also able to transmit real-time data via the Internet when processing products, taking into account the role of a cloud platform for storing and analysing big data. Through such a system, intelligent machine tools can also help managers to allocate resources, quickly solve the problem of costing and remote control, and provide efficient allocation of production tasks, product customisation, machine rental and a range of other aspects. The i5M8 series breaks the mould of traditional manufacturing operations by automatically generating machining processes through intelligent programs and transmitting them remotely to the machine platform, allowing mechanical designers to access these complex mechanical parts easily and quickly from home. It is not only manufacturers who benefit from this system, but also managers who, through the i5M8 intelligent system, are connected to intelligent logistics so that users can see the progress of the entire production process: from the designer's drawing to the final product, it is a personalised and "I want it, I get it" scenario.

The "Industry 4.0" of the 5G era will usher in a new situation in which there will be no dead ends in management and strict control of production and logistics nodes, and under this mode of operation, high efficiency in resource allocation will be achieved.

Agricultural Revolution

Agriculture is vital to the survival of mankind and is the foundation of the country's economy, although the prosperity of the big cities is attracting more and more people from the countryside, but people have to eat, and people have to grow food and fruit and vegetables, so the importance of agricultural development for a country cannot be overstated. Traditional agriculture relies on natural conditions and manual work using human labour and agricultural tools. The growth, yield and quality of crops are controlled by the accumulated experience of generations of farmers. Early agriculture aimed at self-sufficiency. The dependence on the heavens and the manual scattering of labour have long been the main features of agriculture. Traditional agricultural production is low, yields are very limited, production methods are traditional and subject to the natural environment. Although this mode of production allows for a balanced supply and demand, the biggest problem facing traditional agriculture is the uncontrollable factors that can affect the harvest in the event of a natural disaster.

Traditional agricultural production and operations rely on past experience and the output of crops depends on individual effort and natural conditions. As human society has progressed and technological revolutions have emerged, traditional agriculture has evolved from the primitive model of subsistence and manual labour to modern agriculture, incorporating and using new machinery and technology. As human society has progressed and technological revolutions have emerged, traditional agriculture has evolved from the primitive model of subsistence and manual labour to modern agriculture, incorporating and using new machinery and technology. In the age of big data, agronomists can find the best ratios for growing agricultural products by integrating historical data such as humidity, soil quality, air indicators and weather forecasts, and analysing data on crops and other relevant factors to significantly improve the yield and quality of agricultural products.

"Technology for Agriculture" will no longer be a slogan in the future 5G era, when big data is combined with agriculture, modern agriculture will be upgraded once again, showing us a beautiful picture.

The Earth Can Feed More People

While agricultural technology continues to advance, we must also look at a real problem: as the world's population grows each year and land resources dwindle, the exploitation of existing resources and land reclamation is not a sustainable development strategy. In addition, there are still many barren and infertile areas around the world that cannot be used, and there are still areas that are relatively poor in agriculture.

In the case of China, with the accelerated urbanisation process, agricultural land resources are decreasing, and the arable land reserve resources are not only few in number, but also mostly scattered and less available due to poor natural conditions. There is also a lot of misuse of existing agricultural land, and the phenomenon of abandonment of arable land is serious. In addition, the quality of land in our countryside is now clearly deteriorating and its sustainable use is not high. To realise the new agricultural model, the use of land resources must be combined with the latest technology and big data to improve the efficiency of use, so that the limited agricultural land resources can feed more people.

In the early 1980s, the United States pioneered the concept of precision farming, using data to manage farm operations and developing and equipping agricultural tools that could carry data. The idea of precision farming was put on hold because the hardware was not yet mature enough and the smart network was not yet in place. Ten years later, specialised agricultural data companies have emerged in the USA to produce intelligent agricultural equipment. These devices have an intelligent learning function that allows real-time analysis of weather changes and allows farm operations to be adjusted accordingly.

In a large area of farmland, each plot may have different moisture indices, nutrient indicators and crop growth. Traditional farm management relies heavily on human labour, with a certain number of farmers assigned to a certain area to carry out tasks such as sowing and fertilising, and they do not distinguish between the differences in land on the farm, sowing the same varieties at equal intervals. This approach often results in a certain percentage of crops growing poorly, which is a huge waste of land resources. Precision agriculture overturns the traditional use of land resources. Professional automatic seeding machines with soil analysis function, based on scientific analysis data, sow intensively on fertile soil and thinly on land with low fertility. In addition to analysing the soil, the smart device also carries out seed analysis to find the optimum ratio of seed to soil and to change seed varieties. Large areas are sown using these automatic seeders with powerful intelligence, which can be managed efficiently by just one person and, thanks to precise sowing, can increase yields per hectare.

The intelligent seeder has greatly improved the quality of the work on the farm, increasing the single seeding rate to 99%, and the entire workflow can be monitored in real time. With real-time data recording, farm managers can judge the operation of the machine based on the data and can stop the machine at any time to correct any abnormalities in the data, effectively preventing unnecessary losses. In addition to intelligent seeders, other agricultural equipment is also used throughout the agricultural activity. Through accurate data analysis, we are able to accurately control the proportion of all raw materials, maximise cost savings and make the most of every inch of land, which was previously unthinkable.

Rodney Schilling, a farmer who lives on a farm in Illinois, USA, runs a field of about 7900 acres with his father. The father and son are the only two people on the farm and no workers are employed, even at the busiest times of the year, Cillin also only needs the smart farm machinery and a tablet computer to do all the work on the farm easily and efficiently. The machines are equipped with a satellite navigation system and an autopilot function that allows him to do whatever he likes in the cab, while the machines work on a set route and the entire progress of the work is recorded at all times and can be monitored by Cillin remotely. He is not the only one who manages his farm in this way, and there are more and more farmers like Celine in the United States.

If every inch of land can be developed according to the concept of precision agriculture, then in the future 5G era, agriculture, like industry, will be organically integrated with the information industry to achieve big data operations. Such smart farms will create greater yields, save more money and truly achieve the goal of feeding more people with limited land resources.

Relying on the Harvest to Become Controllable

The main characteristic of traditional agriculture is that it depends on the weather and is influenced by natural conditions. With the development of human society, the progress of agricultural technology and the reduction of agricultural population, science and technology has become the main direction of agricultural development, and the transformation of agriculture from traditional to modern is an inevitable trend.

Israel, a small country, is a successful example of how science and technology can contribute to agriculture, and its miracles are particularly remarkable.

Israel covers an area of 14,900 square kilometres, two-thirds of which is desert and mountainous, with an average annual precipitation of around 200 millimetres and less than 3% of the world average in water resources per capita. Such a natural environment is too bad for agricultural development. However, a country with extremely poor land resources and a severe lack of water resources has become a globally renowned agricultural powerhouse through the use of science and technology to promote agriculture.

Throughout the mythology of Israeli agriculture, there have been three broad stages of development:

In the early 1950s, the Israeli government began a major effort to develop agriculture. At a time when the country was at war and under great economic pressure, agriculture became a lifeline for Israel's economic recovery. The government supported the reclamation of land and the establishment of settlements throughout the country with the aim of achieving self-sufficiency in food. In 1952, the Israeli government actively introduced cotton to clothe the nation and to export it abroad. A year later, Israel began to develop the desert and implement the "North-South Water Transfer" project.

Beginning in 1952, the Israeli government spent 11 years and invested $150 million to build the "North-South" water pipeline, but traditional agricultural irrigation techniques were difficult to adapt to advance the transformation of the desert, and progress on the project was very slow. In 1962, a farmer accidentally noticed that crops were growing well where a water pipe was leaking, and it turned out that the water was seeping into the soil in the same place, which not only effectively reduced evaporation, but also provided good control of fertilisers and pesticides. This discovery was strongly supported by the government and two years later the famous Nettlefim drip irrigation company was founded. Since the 1960s, drip irrigation has rapidly revolutionised agriculture in Israel, resulting in a rapid increase in agricultural production, the transformation of the desert, the expansion of cultivated areas, and the transformation of desert cities into green and new. Since then, Israel has continued to develop new drip irrigation technologies and improve drip irrigation equipment, fundamentally changing the traditional agricultural model. Drip irrigation technology saves 1/3~1/2 of water than diffusion irrigation, increases the yield per unit area of land by more than 1/3, and the utilisation rate of water resources is up to 90%. In Israel today, the widespread adoption of drip irrigation systems, which are technology-based and computerised, and the export of this technology to other countries around the world, has greatly alleviated the water crisis.

Since the 1980s, Israel has been actively pursuing the industrialisation of agriculture, implementing a strategy of structural transformation of the agricultural industry and, at the same time, opening up international markets and establishing a modern agricultural management system that integrates agricultural technology and factory management. The range of agricultural products has also expanded from the original single grain to include high quality meat, flowers, vegetables and fruits, which are exported overseas. Science and technology have become a national policy in Israel, with the state encouraging farmers to learn new technologies and strongly supporting agricultural research. Israel now has more than 3500 high-tech companies, seven research institutes, more than 250 doctors and researchers working on more than 750 research projects and hundreds of millions of dollars in government funding for agricultural research every year. Today, Israel is an innovative country that has not only changed agriculture from a dependency on the heavens to a controlled one, but also attracts a large number of agricultural experts from around the world every year.

Smart agriculture in the 5G era, which applies the Internet of Everything and big data to modern agriculture, is expected to be free from natural constraints to a greater extent. Big data will enable the systematic analysis of wastelands and areas with poor land resources, the effective scientific use of these resources, and the use of precise analysis of climatic conditions to tailor a plan for the exploitation of a particular piece of land, thus increasing agricultural production.

Factory-Based Agricultural Production

Nowadays, there is a trend to make full use of big data to grow crops on a large scale. With the help of technology, land can be maximised and the future of agriculture will be factory-based. In fact, the industrialisation of agriculture in Israel is now taking shape in the form of production plants.

The United States is a highly developed agricultural powerhouse and has been open to agricultural data for a long time and has focused more on collecting agricultural data. It is conceivable that once we enter the 5G era, the US will be committed to using big data to systematise weather, soil, seeds, fertilisers, crop chemicals, etc. to create a unified model and make this system available to farmers and suppliers to interoperate information, flatten processes and maximise agricultural yields and profits. In addition, manufacturers of agricultural machinery will also become part of the whole new agricultural chain. As the Internet of Everything matures in the 5G era, weather stations, traders, technology providers and related partners will also join the entire value chain of agricultural development, sharing information and enabling smart factory-based precision agriculture.

Of course, it is not only the United States that will use this new model of agricultural operation, but also other countries that are actively using big data to explore the development of smart agricultural chains. But thanks to exceptionally advanced technology, smart farming in the US is a little ahead of other countries: there are a number of agricultural data companies, including hardware and software from Precision Sowing, weather data analysis products from Climate Group, SaaS predictive data systems from Smart Farm, and historical weather data location developed by Farm Logs. Monsanto, which was founded in 1901, is very optimistic about the future of data and agriculture, and has acquired a number of data analytics organisations in an attempt to continue its dominance in the agricultural sector.

With the advent of smart farming equipment and the rise of the data farming analytics industry, the future of farming will be able to operate like an industry and easily achieve large-scale operations. When the production and processing of agricultural products are monitored in real time, the safety of agricultural products, food and foodstuffs will be reconstructed.

Food Security Addressed Once and for All

Food safety has always been a major concern for countries around the world because of the health of the population. In China, in particular, major food safety incidents such as melamine milk powder have come to light in recent years, causing panic among the public and devastating the companies involved. In the future 5G era, the problem of food security will be completely solved under the monitoring of big data flows such as smart agriculture, smart IOT and smart logistics.

The World Health Organization has proposed a big data approach to support food safety decision-making and has established a food safety platform, Foscollab, which integrates data from different disciplines and enterprises, including agricultural, food and public health indicators. In the future, when big data is mature, the authorities can collect data on food safety through online databases, the Internet, mobile smart terminals and social media. With the advent of mobile communications to 5G networks, it is expected that smartphones will be able to monitor food sensitively and synchronise production records to computers and official food data centres, thus bringing food safety under the scrutiny of the whole population and creating good regulation.

The application of big data to food security systems, in addition to data collection and storage, allows visualisation tools to provide pictures and make location correlations, and even in the initial analysis of information on environmental factors, pathogens and sources of contamination can be predicted and unsafe factors can be eliminated in advance. In addition to this, the detection function of big data can also be used to respond to unexpected food safety incidents. In 2011, the "enterohaemorrhagic Escherichia coli" incident in Germany, information on the presence of these bacteria was collected in time in different regions and professional detectors used this data to screen each household for secondary infections by testing them, providing rapid preventive measures and ultimately preventing the incident from worsening.

In the age of the intelligent Internet of Things, where everything is connected and food safety is linked to data, the food security system will become even better. It is very likely that people will be able to see a very detailed safety monitoring database on their mobile phone when they buy a particular food or ingredient, and the data will be updated in real time.

4

Human Society in the Post-5G Era

5G Accelerates the Internet of Everything on the Ground

Human society has undergone six information revolutions.

The first information revolution: the invention of language, which allowed information to be shared and helped apes evolve into humans.

The second information revolution: the invention of writing, which made it possible to record information, was a sign of the emergence of human civilisation.

The fourth information revolution: the invention of radio, which allowed information to be transmitted in real time over long distances, was an important symbol of the political, military and cultural impact of modern history.

The fifth information revolution: the invention of television, which made it possible to transmit information in real time over long distances, thus bringing warmth and emotion to information. Television is an important information platform that has influenced human society in the last hundred years. Its invention is an important milestone in the development of human culture and entertainment.

L. Xiang, *The 5G Era*, https://doi.org/10.1007/978-981-99-4534-4_4

The sixth information revolution: the invention of the Internet has opened up a great revolution in human information transmission by making it possible for information to be transmitted over long distances and in real time with multimedia interaction in both directions. The emergence of the Internet has challenged the media, entertainment and social management, and has largely influenced every aspect of global politics, economics, culture and entertainment, reshaping human society.

The basic idea of the Internet was to allow the free and fast transmission of information, and the basic information carried on the Internet was textual information. Freedom, openness and sharing have become the spiritual core of the traditional Internet. In the early days of the Internet, there was a famous saying that was widely disseminated and even enshrined as a basic tenet: "On the Internet, no one knows you're a dog". At that time, the Internet did not require a real name and shocking speech was not regulated, which was the basic value of the Internet.

In terms of technology, the traditional Internet uses the TCP/IP protocol, based on the IPv4 system, to build a simple network that facilitates the transmission of information and has great scalability. However, this network system lacks a management and security layer from the outset and is vulnerable to attacks and is poorly secured. Information leaks, security attacks, phishing and Trojan horse software are everywhere, and Internet fraud is rife. Such a network is clearly not capable of performing the tasks that require security. As a network that is primarily responsible for the transmission of information, it is also more difficult to take on additional services.

What is the seventh information revolution of the future? This question is now in front of humanity.

The first six information revolutions in human history have solved various problems of our information transmission, making information sharing, recording, long-distance transmission, as well as long-distance real-time transmission, long-distance real-time multimedia transmission, long-distance real-time multimedia two-way interactive transmission a reality. In the age of the Internet, man has solved essentially all the problems of transmission.

In the seventh information revolution, mankind is moving from the age of transmission to the age of sensing. What else does humanity need for the next information revolution? Obviously, it is in the case of a good solution to information transmission, extending human organs, so that the senses help us to understand more about the unknown world, and at the same time collect, organise, process these data, become big data, and constantly through intelligent learning, and finally form a service. Such a huge system is far from being carried by the traditional Internet, and is not only an extension and development of the traditional Internet, it should be a revolution on the basis of the traditional Internet, or a new reconstruction, which includes the reconstruction of several capabilities.

The seventh information revolution is the Smart Internet, which is a new capability consisting of mobile connectivity, intelligent sensing, big data and intelligent learning. It not only solves the transmission problem, but also has sensing functions, while big data and intelligent learning can collect, process and integrate data, and provide intelligent services on this basis.

In this system, 5G is the foundation of the seventh information revolution and the fundamental guarantee for the completion of the construction of the seventh information revolution.

Some may ask: 1G to 4G also provides mobile communication capability, and 4G network speed has already reached 100 Mbps, so why do we need 5G?

It should be noted that 5G offers a completely different mobile communication capability to previous generations: the first generation of mobile communication was only capable of voice communication; the second generation mobile communications have the ability to communicate with data, but at very low speeds; it is the third generation of mobile communications that has taken us from the voice era to the data era in earnest; the fourth generation of mobile communications has brought mobile communications to a new level through increased speed, on the basis of which mobile payment, e-commerce, bike-sharing and car-sharing services have quickly developed, greatly improving social efficiency and enhancing the user experience.

However, it is clear that the 4G network is not adapted to the development of the smart Internet, and there are many problems that it cannot solve, apart from the high network speed.

For 5G, 3GPP has defined three major scenarios, and on the basis of these three scenarios, we see that mankind has even higher expectations for the future of mobile communications.

5G has six main features: high speed, ubiquitous network, low power consumption, low latency, interconnection of everything, and reconfigured security. These six features show that 5G will not only enable networks with higher speeds, far exceeding 4G, but more importantly, will be able to cover every corner of social life and provide services anywhere, anytime. The ubiquitous network is the basis for service assurance and quality improvement, as well as a strong support for the development of new services. Currently, China is the best developed country among the world's major powers in terms of mobile payments, and the ubiquitous network plays an important role in this. In today's China, not only in cities, but also in remote rural areas, payment can be made at any time anywhere there is an Internet connection, making it possible for mobile electronic payments to develop at a rapid pace. An interesting phenomenon is that countries like the United States, which are economically and technologically advanced, are more innovative, but mobile electronic payments have not developed, and one important reason is that the network coverage is not good enough, users cannot pay with mobile once, next time they will use credit cards.

Low power consumption is even more unattainable with 4G. There are a large number of IoT applications that cannot be deployed without low power consumption. It is not possible to carry a battery with you at all times and recharge it every day, as is the case with many businesses and applications. For the Internet of Things to be connected, low-cost modules and low-power networks are required. One of the basic requirements of 5G is to support large-scale IoT applications, so in addition to high speed, there is also eMTC to support medium-rate access, and NB-IoT to support low-rate access, thus achieving low power consumption. Although the deployment of NB-IoT can be built on 2G networks, this standard was proposed under the demand of 5G and the network construction started with the arrival of 5G.

Low latency is also something that pre-5G networks could not achieve. In scenarios such as unmanned driving and industrial control, which require high accuracy, the delay of 20 to 80 milliseconds in the 4G era is obviously not enough to meet the needs, and reducing the delay to 10 milliseconds or even 1 millisecond will require a large-scale transformation of the network, as well as the introduction of edge computing and many other technologies into the construction of the 5G network. The low latency will allow the network to be controlled with much greater accuracy in order to do work such as drone formation, and the application scenarios for 5G will be greatly increased.

Today's mobile networks are very limited in the number of terminals they can support, with no more than 500 mobile phones connected to a single base station and a similarly limited number of terminals accessible within a small sector. The Internet of Things requires that in addition to mobile phones, a large number of social public management and daily life products can be networked, including cars, charging piles, parking spaces, utility poles, street lights, manhole covers, monitors, traffic lights, door locks, air purifiers, air conditioners, range hoods, heating valves, electric lights, refrigerators, washing machines, rice cookers, sockets, glasses, belts, shoes, bracelets, watches and other social public services, smart home products and personal household goods can be networked, the vision is to have one million equipment terminals per square kilometre networked, which requires airports can support a large number of devices to form the ability to interconnect everything.

Another important feature of 5G is the reconstruction of the Internet security mechanism. The security mechanism of the traditional Internet is very weak, 5G is not an information transmission platform, it has to penetrate into every field of social life, including public administration, intelligent transportation, intelligent home, intelligent health management, industrial Internet, intelligent logistics, intelligent agriculture, etc. A large amount of data involves national security, public security, personal privacy. If this network, like the traditional Internet, is easily breached, it will result in a large amount of information being leaked or even managed and controlled by hackers, so a network without security would rather not be built, as the damage and impact is too great. For the Smart Internet to operate securely, security mechanisms must be re-established, and it may even be possible to consider reconstructing a security-first network outside of the traditional Internet.

5G, with its six features, solves the problems of speed, ubiquity, power consumption, latency, interconnection of everything, security and other aspects for mobile connectivity, allowing a large number of sensors to be deployed, turning smart sensing from a concept into a product with more functions, entering the field of public administration and the daily life of ordinary people, and only on the basis of smart sensing will there be smart learning and the formation of valuable services.

The concept of Artificial Intelligence has been around for 60 years, the concept of Internet of Things has been around for more than a decade, but it is difficult to turn it into a valuable service. One of the major reasons is that it is costly and complex to deploy, and cannot really enter the general public administration and the lives of ordinary people. Low-cost, efficient and capable 5G communication services will be a solid foundation for the Smart Internet and an important driving force for its development.

The deployment of 5G will also significantly reduce the cost of communications and facilitate business development and greater consumer adoption. In the 2G era, 1 GB of traffic cost the operator around 10,000 RMB, and with such high prices, 5 MB of traffic per month was the norm, and 30 MB of traffic was considered high; in the 3G era, with the increase in data capacity of communication networks, more data traffic can be provided, with the price of 1 GB of traffic dropping to around RMB 500 and users using around 100 MB of traffic per month; in the 4G era, the price of 1 GB of traffic has dropped to less than RMB30, or even about RMB10 in some cases, and users are using more than 1.5 GB of traffic per month, which is a significant increase from the past. With the fall in traffic prices, a large number of video, social, e-commerce and e-payment services are available to users at will, making social services much more efficient; the arrival of 5G, traffic charges will drop to 1 GB 1 yuan, or even lower prices, which requires the consumption of more traffic AI, VR, video business will bring a huge boost.

The intelligent Internet service system formed by mobile Internet, intelligent sensing, big data and intelligent learning will bring us from the era of information transmission to the era of sensing, which will greatly extend the human organ, and the artificial intelligence will greatly enhance the human sensing ability and form more efficient services on this basis. This will usher in another new era of the information revolution.

6G a Technology Evolution, Not a Revolution

After 5G, 6G technology is already being studied by existing countries. Compared to the revolutionary changes of 5G, 6G should be a refinement, enhancement and further optimisation of 5G technology.

This is the same as the evolution from 3G to 4G. 3G is the move from the era of digital communications to the era of data communications. 2G also has digital communication capabilities, but mainly for texting and caller ID, Internet access is a very minor capability and the amount of data used by users is very small, the core network and billing system of telecom operators are based on voice. After entering 3G, although voice is still very important, the network is no longer a voice-centric network, data services have become important, and the core network and billing system is a data-centric system. The transition from 2G to 3G is a revolution for the communications network, a completely new transition.

However, while the 3G network is objectively slower, 4G has greatly improved the network speed and made the user experience much better. We can think of 4G as a technological evolution, just an increase in network speed, with no revolutionary change in its infrastructure and basic capabilities compared to 3G. Even a large number of 4G services have been built up gradually on the basis of 3G. It is only with the arrival of 4G that the efficiency of these services has been greatly enhanced, ushering in an explosive period.

5G puts more demands on the technology, with faster speeds and lower power consumption and latency. Compared to the 4G network,

5G will involve a restructuring and rebuilding of the entire network structure, as well as significant changes to the core network, management and billing system. The services formed on the basis of 5G will no longer be limited to the transmission of information, but will be a smart Internet, a network that will be integrated by mobile connectivity, intelligent sensing, big data and intelligent learning to form new capabilities, which can be described as a revolutionary change. The 5G revolution represents a dramatic change in technology, management, billing, business models, commercial models and business formats.

So, will 5G make all the visions come true and achieve the perfect feel and experience? Obviously not. 5G will also undergo a long-term evolutionary process, gradually improving from the network, management system and other aspects of continuous improvement, so in this sense, 6G will not be another technological revolution, revolutionary changes in the system, but on the basis of 5G, according to the actual operation, to find new problems, or for the shortcomings of 5G to improve.

6G Will Allow Further Network Convergence

At present, the mobile and satellite networks are two separate networks, each operating separately and not playing a complementary role to each other. 6G is likely to see a trend towards the integration of heaven and earth, with a network that can not only provide terrestrial coverage through terrestrial base stations, but also through low-orbiting satellites and high-orbiting satellites, and deep coverage on the ground through numerous repeater stations and small base stations. The fusion of networks, on the other hand, allows for a balance between the surface and the point, with high and low orbiting satellites, ensuring that the surface, in any corner of the planet, may have a network presence; at the same time, deep coverage can be provided in densely populated areas or areas where the current network is not dense enough, and can go deep underground, such as subways, tunnels, car parks, mines and other special places where the signal is poor.

In addition to being covered at a certain depth below the surface, underwater communication should also be possible in the 6G era and be part of the overall network coverage system. For example, in offshore, rivers and lakes, water quality, underwater plants, underwater organisms, underwater aquaculture products, changes in water temperature, nutrient content in water, pollutant content, underwater dam displacement and many other areas need to be monitored, and underwater monitoring with network coverage will largely enhance monitoring and management capabilities. 6G can also play an important role in collecting ocean-related data, monitoring environmental pollution, climate change, undersea anomalies, seismic and volcanic activity, detecting targets on the seabed, as well as transmitting images over long distances underwater, and even in the military. It is clear that in the 4G and 5G era, underwater network coverage was not even considered, and that a breakthrough should be planned and achieved in the 6G era.

Specifically, underwater wireless communication can use a variety of technologies such as underwater electromagnetic wave communication, hydroacoustic communication and underwater quantum communication. Radiofrequency (RF), a high frequency electromagnetic wave with frequencies above 100 KHz and capable of radiating into space, will have a greater opportunity in underwater wireless communication. The improved technical capabilities make it possible to achieve high speed data transmission of 100 Kbps or more, as well as to resist the effects of noise, achieve relatively low latency and low power consumption, and have a high level of security. RF communication has the potential to become an important option for underwater wireless communication. In addition, underwater laser communication and underwater neutrino communication will also become technologies of great interest, but these technologies need to be improved.

An important step forward for 6G is to make the network truly ubiquitous through technologies and methods such as satellites, ground stations, small base stations and underwater base stations, which will link the sky, ground, underground and underwater into one whole. There are still challenges for the communications industry to realise these capabilities, and there are many areas that must be improved and refined if these networks are to be linked together to form new business models.

Adopting More Spectrum to Increase Efficiency and Capacity

In order to achieve greater bandwidth, more spectrum must be used for mobile communications. In the 5G era, 800/900 MHz is being used for IoT bands, and 3.4 GHz–3.6 GHz and 20 GHz–60 GHz spectrum are being considered for 5G, thus greatly increasing the available spectrum and improving network capacity and bandwidth.

6G network needs more bandwidth, the peak speed of 6G will reach 100 Gbps, while the peak speed of 5G is only 20 Gbps, and the single channel bandwidth will also reach 1 GHz, while the single channel bandwidth of 5G is only 100 MHz, by combining multiple 1 GHz bandwidths, the final speed of 100 Gbps can be achieved. To achieve a bandwidth of 1 GHz for a single channel, the spectrum in the lower bands is clearly not enough to support it, which means that more spectrum has to be taken out for mobile communications. Millimetre waves have already been adopted in the 5G era and it can be expected that terahertz waves will enter the picture in the 6G era.

Terahertz waves have a wavelength of 3 μm to 1000 μm and a frequency of 0.1 THz to 10 THz, and are electromagnetic waves between microwaves and light waves, combining the advantages of microwave communication and light wave communication, which to some extent gives it and the rest of the electromagnetic wave different characteristics, namely high transmission speed, high capacity, strong directionality, high security and strong penetration.

It is a basic principle in the field of communications that the higher the frequency, the greater the capacity. Terahertz waves, which are 1 to 4 orders of magnitude higher in frequency than the microwaves currently in use, can provide wireless transmission speeds of more than 10 Gbps a height unattainable by microwaves and are of greater benefit in solving the problem of bandwidth-constrained information transmission.

Terahertz waves are obviously difficult to use for long-distance transmission, but today's communication networks are basically backbone transmission networks built from optical networks, with base stations to extend the optical network to form a mobile network, users can switch between different base stations at will, so that they do not drop the line at any time and can support high speeds. The mobile network of the future will be a wide-area cellular network, requiring long-range and wide-area coverage, with smaller and smaller cells. The network of the future will be an ultra-high density, ultra-high speed, densely deployed network system with many small base stations, mainly in the home, under a huge optical network. In this system, where high speed transmission is required in densely populated areas, terahertz technology can take advantage of high speed, high capacity and high penetration to enable deployment in office and home environments.

The higher the frequency band of electromagnetic waves, the poorer the ability to bypass and penetrate, millimetre waves are difficult to penetrate, so in the city, whether in the office or home environment deployment is a major problem, while terahertz technology because of the proximity of light waves, has better penetration, while it does not need to compete with other industries for spectrum, so you can achieve a large bandwidth. It is because of these characteristics that terahertz technology is particularly suitable for deployment in office and home environments in densely populated urban areas where distance is not an issue, but bandwidth and penetration are.

Smarter Communication Networks

In the future, AI functions will increasingly be introduced into communication networks, turning them into an intelligent network.

In the traditional communication network, the trinity of pipe, management and business is provided by the telecom operator, but the telecom operator only carries one kind of business, voice, and the management is also very simple. In the 5G era, networks will become even more complex. If we were to build multiple communication networks, i.e. multiple pipeline systems, not only would this increase social costs significantly, but social resources would also be insufficient. To give a simple example. Cities have many systems such as electricity, telecommunications and sewage, all of which require pipelines. If each system built a pipeline of its own, it would not only be costly but also very difficult to manage, which would require a pipeline to share resources.

With the development of 5G, the communication network is no longer just a network for voice communication between human beings, nor is it just a network that can provide Internet access services, but will also carry a large number of Internet of Things services, city management services, intelligent transportation services, smart home services, and the security, priority and resource allocation of different services are all different. Such a network must have a strong and intelligent management model to intelligently manage the different services and users. For this reason, AI technology must be integrated into the network to identify different users, their behaviour and terminals, and to allocate resources and manage tariffs on this basis.

A question worth exploring is whether the future 6G network, which is a complex system of networks with extremely high resource requirements, will require several networks to be set up in one country. On the one hand, resources are very limited, and on the other hand, interference between networks is a major problem. When the network reaches every corner of society, if one corner is covered by three or more networks, not only is it a waste of resources, but the quality of the network also suffers.

For this reason, the 6G era should explore the construction of only one full-coverage network in a country. In this way, not only can we make full use of resources and reduce the waste of allocating several sets of equipment in each location, but we can also greatly reduce the cost of renting station sites and avoid interference between networks. Such a network will be of higher quality, take up fewer resources and have less impact on users.

Of course, it is not possible to merge multiple operators into a single operator with no competition at all, which would require a separation of networks and services: multiple operators using the same network to provide their own services. The competitiveness of telecom operators is mainly reflected in the management, platform, research and development, operation, service and support capabilities, because their respective capabilities are different, the level of service provided is different, so that users have different choices.

Telecom operators establish their own service management platform, billing platform, charging platform, service support platform and customer service platform on the network, and then provide different products and services to individual users, enterprise users and institutional users. Users can identify themselves with one ID and use many different services at the same time, without having to register an ID for each service, i.e. one ID, many services, one fee.

The network can also use AI technology to identify users and then provide different network resources and security guarantees according to different user identities and endpoints.

At that time, there will be only one communication network in a country, and multiple network capabilities will be superimposed at multiple levels, so operators will no longer be competing for network capabilities, but for R&D, management, operation and service capabilities.

Where the Future of Information Transmission Lies

After 5G and 6G, human mobile communication technology will continue to evolve. So, what direction will human information transmission take in the future?

Communication Networks Need to Extend into the Far Reaches of the Universe

When the environment currently accessible to mankind—in the sky, on the ground, underground and in water—is covered by a network, the human information and communication network will extend to the far reaches of space. After 6G, mankind's reach should not be limited to the Earth, deeper exploration of the universe is a subject that must be faced. By then, humans will not only be able to land on the Moon and Mars for scientific research and even long-term stays, but may also be able to reach more distant planets.

Human communication after 6G must be oriented towards the distant universe, and on a planet as close to Earth as Mars, a communication network should be set up to transmit information to and from Earth. The establishment of ground stations on the Moon or Mars, or the relay of multiple communication satellites to cover the surface of the planet, builds up a communication system between the planet and the Earth.

Only when a communication system is established between the Earth and a planet will that planet be integrated into the Earth's life system and it will be possible to study it systematically and to really understand what is happening on that planet. With communication systems, humans on Earth will be able to communicate with other planets, people from different planets will live in one system, and humans from far away in space will no longer be alone.

The earliest established Earth-to-earth communications with other planets used ultra-longwave and longwave, which allowed for better long-range transmission and were relatively easy to establish. However, such a network transmits a small amount of information and has a small bandwidth, which does not really allow for the transmission of large amounts of information. This requires the use of UHF centimetre waves for network deployment and the relaying of signals via multiple satellites, ultimately allowing for larger bandwidth transmissions. An Earth-centred communications network could be established on the near-Earth planet that could support initial text and voice communications and eventually high speed, high-definition video communications.

The formation of an efficient communication network, similar to that of a ground station, linking multiple planets in the universe through multiple satellites, will allow a large number of scientific research activities to be transformed from single, discontinuous observational studies, which are difficult to monitor comprehensively, to continuous, comprehensive and systematic studies. However, not enough is known about the vastness of the universe, and a more extensive cosmic communication network would be an important step forward for mankind. The long-term goal is to conduct scientific research and study on other planets and to migrate humans, and to achieve these ambitious goals, the establishment of a smooth communication link between the planets and Earth is a fundamental requirement.

Such a communication network should not be a single point, occasional, discontinuous communication, nor should it have low speed and little information, for which reason it is necessary to consider not only what technology to use, whether radio waves, light waves, or other media, but also how to meet the requirements of long distance, high speed and anti-jamming cosmic communication. Consideration could be given to establishing ground stations on near-Earth planets such as the Moon and Mars for communications coverage and large communications repeater stations to provide stable network support.

In addition to the transmission technologies used to ensure long-distance transmission, large bandwidths and high speeds, resistance to all kinds of interference in the universe and high security, the supply of energy is also taken into account. The question of how to provide energy support for the establishment of terrestrial communications on other planets is a major issue that is still to be solved as soon as the technology is developed.

In any case, mankind will gradually extend its communication network from Earth into the universe. The space in which humans live will also expand from Earth to other planets further afield.

Human Communication to Break the Spectrum Bottleneck

Human beings have long had to rely on radio for the transmission of information for mobile communications. The long transmission range of ultra-long wave (VLF) and its high penetration capability allow for communication between submarines and shore, navigation at sea, etc. Long wave (low frequency) can transmit over long distances and has a high penetration and bypass capability, allowing communication at moderate distances within the atmosphere, including underground rock communication and navigation at sea. Medium wave (IF) is widely used for broadcasting and maritime navigation. Shortwave (HF) is used for long-range shortwave communications and shortwave broadcasting. UHF (Very Short Wave) has a longer transmission range and a much larger bandwidth and is used in a variety of communication modes, such as: ionospheric scattering communication (30 MHz–60 MHz), meteor afterimage communication (30 MHz–100 MHz), artificial ionospheric communication (30 MHz–144 MHz), communication to space flying bodies (aircraft, missiles, satellites) inside and outside the atmosphere, communication to television, radar, navigation and mobile communication inside the atmosphere. The decimetre wave (UHF) has a small transmission distance and weak penetration and bypass capability, and is mainly used for tropospheric work-scattering communications (700 MHz–1000 MHz), small-capacity (8–12-way) microwave relay communications (352 MHz–420 MHz), medium-capacity (120-way) microwave relay communications (1700 MHz–2400 MHz), while 4G and 5G mobile communications are also widely used in this frequency band. Centimetre wave (UHF) has a larger bandwidth but poorer penetration capability and is used for high-capacity (2500, 6000) microwave relay communications (3600 MHz to 4200 MHz, 5850 MHz to 8500 MHz), digital communications, satellite communications and waveguide communications. Millimetre waves (very high frequencies) have greater bandwidth, but transmit closer and have poor penetration and are used for communication when penetrating the atmosphere. In order to obtain greater bandwidth, millimetre waves, previously thought to be impossible for mobile communications, are now also attracting attention for close range, high-speed mobile communications.

For communications, spectrum is always a bottleneck that cannot be broken. To communicate, spectrum resources must be used, and valuable spectrum resources are limited. Breaking the limits of the spectrum and finding alternative media is a viable way forward. We could expand the spectrum to terahertz, but its resources would still be limited and, more importantly, it would probably be subject to too many external conditions and restrictions.

In ancient times, the most advanced communication system was the post, which could transmit information thousands of miles away in a day through a post station located tens of kilometres away, but such a large and efficient system was only available to the powerful. The technology was not available to establish low-cost, more efficient long-distance communications, and it was difficult to imagine how information could be delivered thousands of miles away without a one-stop relay.

In modern times, with the development of communication technology, mankind discovered the existence of radio, which could travel thousands of miles and at a very low cost, without the need for a relay station. Of course, for those who worked in communications in ancient times, radio transmission could not turn messages into written letters and was of little use until the advent of coding technology, which made it possible for radio waves to carry text thousands of miles away. Since then, the telegraph, telephone, radio and the Internet have completely overturned the function of ancient communications, and today's communications have long since moved away from the stagecoach model and do not require the physical transmission of information.

For those working in communications today, it is either radio or light waves that are used to communicate. Light and electric waves are the most efficient tools for transmitting information over long distances, so it is unthinkable to get rid of them.

In ancient communication systems, we could only make horses go faster in order to increase the speed of information transmission. Even with the subsequent change of transport to cars, trains and even planes, the increase in speed was still very limited. Only a change in the medium is revolutionary, allowing the speed of information transmission to change from days, hours and minutes to seconds and milliseconds. There is value in making efficiency gains in the system of light and electric waves, but the resources and speed of this system are limited.

What to do? To break the limits of the spectrum, quantum communication will be the breakthrough.

Optical quantum communication is mainly based on the theory of quantum entangled states, using quantum invisible transmission (transmission) to achieve information transfer. Scientists have experimentally shown that if two particles with an entangled state change, no matter how far apart they are, the other will change instantaneously. Using this property, optical quantum communication can be achieved as follows: the particle with the unknown quantum state is measured jointly with the particle of the sender, and the particle of the receiver changes instantaneously to a state which is symmetrical with the changed state of the particle of the sender, then the information of the joint measurement is transmitted to the receiver through the classical channel, and the receiver transforms the collapsed particle according to the received information to obtain the unknown quantum state exactly the same as the sender.

In the process of quantum entanglement, one quantum state can represent both 0 and 1, and seven such states can represent 128 states or 128 numbers at the same time: one such optical quantum communication transmission is equivalent to 128 classical communications, which is astonishingly efficient.

At present, our understanding of quantum communication is still limited to the transmission of quantum keys, while the future of quantum communication can be used for quantum invisible transmission and distribution of quantum entanglement. The term invisible transmission refers to a "complete" transmission of information away from the physical object. Although quantum communication technology is still in its early experimental stages, it is theoretically possible to replace today's light and electric wave communication with a new communication system, just as it is possible to replace a post station with wireless communication.

Establishing a Sensing System between the Human Brain and an External Chip

A major bottleneck in human communication today is that all communication needs to be recognised by the senses, then stored and calculated in the brain, and then judged. This process is relatively inefficient, as it begins with perception through the sensory system, is transformed into information, is sent through the human nervous system to the brain for storage, is calculated and judged, and is then sent through the nervous system to the limbs for continuous information perception and response through the limbs. This process, which requires multiple conversions, greatly affects the speed of information transmission.

To break through the bottleneck of information transmission, human beings must break through the limitations of the senses and connect a lot of information from the outside world directly with the human brain, eliminating multiple rounds of conversion and transmitting information directly.

It is conceivable to implant a biochip in the human body and connect the biochip to the nervous system of the human brain. Instead of converting text, speech and pictures through the sensory system and forming information, a large amount of information will be sent directly to the human brain and stored on the biochip, achieving a complete integration of carbon-based biological storage and computing and silicon-based storage and computing. This is the ultimate breakthrough in the future of human communication. Instead of converting information through the sensory system, large amounts of information will go straight to the brain, thus reshaping humanity.

Today, the Neuron Chip has been developed as a single-chip system with multiple processors, read/write/read-only memories (RAM and ROM) and communication and I/O interfaces. The read-only memory contains an operating system, the LonTalk protocol and a library of I/O functions. There is non-volatile memory on the chip for configuration data and application programming, both of which can be downloaded via the network. In other words, the neuron chip itself is a memory and at the same time has a communication function. However, it is not yet powerful enough and still needs to be improved and enhanced.

Scientists dream of a future where neuronal chips are "alive", where the organism and the chip become one, where brain cells and silicon circuits become one, and where information stored in brain cells, such as knowledge, dreams and memories, can be freely transferred, copied and extracted between brain cells and the chip. If this is achieved, much of the knowledge and relevant information will no longer have to be learned through a long and complex process, but will be grasped instantly through the transfer of information in the chip.

The transmission of information will then no longer be in the traditional mode of thinking: the perception is carried out by the five human senses, and the information is sent to the brain, where it is analysed, summarised and organised to form knowledge and memory, and then sent to the memory partition of the brain for storage. A considerable amount of information can be stored directly past the senses and can also be searched and recalled by the brain and finally involved in calculation and analysis. This is a substantial change in the way human beings transmit information, with a thousand-fold increase in storage efficiency and a huge impact on human biology, ethics and morality. We will be confronted with the question of whether people are still human, and at the same time the level of intelligence will become more unequal between people, with huge differences between individuals.

In the quest for new ways of transmitting information, scientists have never stopped searching. Today, researchers have been able to mount 16,000 electronic transistors and hundreds of capacitors on a silicon chip the size of a square millimetre, and then use a special protein found in the brain to glue brain cells to the chip, and instead of using this protein as a simple adhesive, the ion channels of the nerve cells are connected to the semiconductor material, so that the electronic signals from the nerve cells can be transmitted to the silicon chip, and then the protein captures the changes in brain waves, converts the information from the brain cells into electronic signals, and interprets them, stores and records the information. The electronic components on the neurochip and the living cells form a communicable link with each other and the electronic signals from the nerve cells are recorded by the transistors on the chip. In the longer term, these recorded electronic signals can be understood and compiled, eventually bridging the information stores of the human brain and electronic information into a complete information system that can be understood by each other. In this way the human being is transformed from a biological being to a fusion of biological and silicon-based human beings. To a certain extent, man began to evolve into a new species.

Energy Storage Breakthroughs to Come

The two most critical capabilities that support an organism are energy and information. While energy is the basis for the existence of a living being, information is the basis for the intelligence of that being.

The history of the evolution of mankind is based on two dimensions, energy and information, which are constantly being enhanced. Energy is obtained through the constant intake of plants and meat, which is the basis for human survival. The long history of human development can be described as a history of human beings doing everything, they can to get their energy back. Whether it was the early days of upright walking, the invention of tools, the formation of group societies or the emergence of nations, one of the key goals of mankind was to gain more energy and occupy as many resources as possible in order to have a long-term chance of survival.

As the industrial age progressed, machines emerged that required energy to operate, and this transitioned from the early biomass to petrochemicals such as coal, natural gas, oil and shale oil. These sources of energy support the operation of machines, which greatly enhance the efficiency of human society and contribute to the development of human civilisation. The struggle for energy has been a trigger for war and political change in recent times in order to ensure one's position.

Nowadays, access to energy is more diversified, with wind, solar, tidal, hydro and nuclear energy… It is well known that bio-energy and fossil energy are relatively limited, but when many energy sources can be converted into electricity, mankind's access to energy becomes unlimited, as many energy sources can be renewable and even inexhaustible, such as wind and solar energy.

In the early days of bioenergy, energy was in physical form, such as food and meat, and its preservation and transportation were a major problem, requiring roads, transport, storage tools and costly human and financial resources. The preservation of meat is even more complicated, as humans have found ways to make cured and frozen products for this purpose. In short, the storage and transport system were then complex and inefficient.

The preservation and transportation of fossil energy sources such as coal, oil and natural gas are also a major problem: coal is bulky; natural gas is not only bulky, but also requires special equipment to transport it without loss; oil is large, requires special equipment for storage and transport, and is also prone to environmental pollution. Manufactured fossil energy products, such as paraffin, diesel and petrol, are very inconvenient to transport and pose a major safety hazard. However, because of their high efficiency, fossil fuels still make up a large part of today's energy mix, but a real problem that has to be faced is that they are not renewable and will gradually be depleted.

The advent of electricity has solved the two major problems of energy transportation and storage. Electricity can be transmitted over long distances through cables, and the vast power grids that allow it to be transmitted in all directions, conveniently, cleanly and safely into homes and workplaces, have been an important force in transforming modern society. Electricity allows products like computers to do all the work of computing and storage. Electricity can also be stored and carried safely and conveniently. Batteries make it possible to transfer energy efficiently, cleanly and easily, so that many tools can have the energy to do their job. For example, mobile phones, laptops and other products that require electrical power can work anywhere at any time thanks to batteries, which are mobile energy storage devices.

The upgrading of energy and the revolution in information are alternating. After gaining enough energy to support them, humans gradually created language for the exchange of information and, having improved their abilities through information exchange, they in turn continued to seek the means to upgrade their energy sources.

Human information transmission has gone through several stages of development, including language, writing, printing, radio, television and the Internet. In addition to transmission, the storage of information is a major problem. For a long time, human information has been stored on paper. The emergence of paper has made a great contribution to human civilisation, it is an important carrier of culture and history, but paper is expensive to produce, does not last long enough and stores very limited information, a book of several hundred thousand words is very large and inconvenient to carry. In order to satisfy people's thirst for knowledge, human beings built libraries to store books and make them available for borrowing. For a long time, libraries were the hubs of information storage and interaction, known as the repositories of human knowledge. However, this model keeps many ordinary people out and the transmission of information remains a major problem.

With the development of human civilisation and the needs of the times, in order to increase the amount of information stored, it was necessary to find a new storage medium that would increase the speed of transmission and allow information to be more easily transmitted and used on a large scale and easily. The advent of silicon has revolutionised the storage and transmission of information in the human world.

Silicon was first discovered in rocks by Lavoisier in 787. Elemental silicon can be found in quartz, agate, flint and common beach stones. Silicon is also a major component in the building materials cement, bricks and glass. Silicon has unexpectedly become an important carrier for information transmission and is the main raw material for most semiconductors and microelectronic chips.

With the advent of computer technology, the storage of information is completely different from the age of paper media, and mankind has entered the age of information explosion. The integrated nature of computer technology is obvious, with electronic engineering, applied physics, mechanical engineering, modern communication technology and mathematics and other disciplines are closely linked, the first general-purpose electronic computer ENIAC is based on the then radar pulse technology, nuclear physics electronic counting technology, communication technology and so on. The development of microelectronics has had a major impact on computer technology and the two have interpenetrated. At the same time, the achievements in applied physics provided the basis for the great development of computer technology.

One of the most important steps in the storage of information by computers is magnetic recording technology, which has unwittingly revolutionised information storage in human history. Compared to paper, magnetic records are much more informative and can store pictures, sounds, images, etc. in addition to text, allowing humans to store information at an unprecedented level. The earliest magnetic recording technology was used for the storage of audio information on records and tapes.

The storage capacity of silicon again greatly exceeds that of magnetic tape, and a small chip can store more information than a library. With the advent of miniaturised computers, intelligent mobile phones and the cloud era, the value of silicon storage also lies in the high speed of transmission that can be achieved.

In the age of paper, there are many “gurus” in society, so-called “gurus”, who are monopolists of information. Information was a scarce resource because there were few stores of information, it was not mobile, and access to information was costly. In this situation, the majority of people do not have access to more information, and the monopolists of information become the masters of knowledge, privileged because of their knowledge, and the respect and admiration of society for the masters becomes a unique cultural phenomenon.

With the advent of magnetic and silicon recording technologies, together with the spread of microcomputers and mobile phones, cloud computing has penetrated every corner of social life and information can flow at high speed and in large quantities. Nowadays, we don't necessarily need libraries to store or exchange information, and many people have not been going to libraries for years. Of course, just because you don't go doesn't mean that you don't learn and read, that you don't receive new information, and that many people get their information through the Internet. Nowadays, if you are willing to take the time to learn, you can find the information you want through the Internet. In the age of the information explosion, with more channels, easier and faster access to information, much lower costs and equal opportunities, the number of "gurus" is rapidly decreasing because there is no longer, or hardly ever, a monopoly on information. In this sense, human civilisation will be greatly accelerated by breaking down the limits of information transmission and by making technology as cost effective and efficient as possible.

More importantly, the emergence of magnetic and silicon storage has also brought about huge changes in human thinking and culture, and laid the foundation for the arrival of big data and artificial intelligence.

Having solved the problem of storing information on a large scale and circulating it at high speed, the storage of energy is also one of the major issues that mankind needs to address.

Today, mankind's access to energy is becoming more and more abundant. Green energy sources such as hydro, wind, solar and tidal are all converted into electricity, but a real problem is that storage is a major problem and grid integration is difficult. Solar energy, for example, can generate electricity during the day when the sun is shining, and the demand for electricity for small-scale general lighting is not large and can be met. But at night, when electricity is needed, there is no sun, so it is difficult to generate electricity. This is also the case with wind and tidal power, which are very unstable and difficult to control precisely.

With the advent of 5G and the smart Internet era, a large number of smart devices require electrical energy to drive them. These devices have moved out of their fixed locations and need to be used on the move, such as smart cars and other vehicles, while a large number of mobile phones, tablets and IoT devices cannot be changed or recharged very often and therefore need to be supplied with energy for a long time.

In the 5G era, mobile phones are getting faster and faster Internet access and bigger and bigger screens, but if they need to be recharged every day or several times a day, this will make the experience much less enjoyable and inconvenient. Many IoT products require long working hours or standby time and are not suitable for daily charging. These require batteries with a larger storage capacity to change this dilemma.

The development of batteries has gone through several stages such as carbon batteries, alkaline zinc-manganese batteries, rechargeable NiMH batteries and lithium batteries. With the advancement of technology, the development of batteries continues to develop towards high density, large capacity, small volume and flexibility, and the rapid popularity of smartphones today is inseparable from the development of battery technology. In other words, without advances in battery technology, mobile communications would have been largely impossible. However, a dilemma is that lithium batteries have reached their limit and it is difficult to increase their capacity to meet more demands in the era of 5G and artificial intelligence, therefore, mankind needs to find new materials and develop ultra-high density batteries to solve the problem of mass energy storage.

From a technical point of view, the core of the solution to mass energy storage is materials. The value of graphene was first noticed when no better material for high-density energy storage than lithium had yet been discovered. Graphene was introduced in 2004 and is one of the thinnest, strongest and most thermally conductive new nanomaterials known, it is 200,000 times thicker than a hair and 200 times stronger than steel, making it the "king of new materials". Graphene has a number of high quality features: robustness and wear resistance, excellent thermal conductivity, good electrical conductivity, resistance to high and low temperatures, and the ability to work in environments ranging from -30 °C to 80 °C. Although graphene does not help to increase energy density, it can significantly increase the speed of charging, making it possible for smart cars to "drive 1,000 km in 10 minutes on a charge". In other words, there is a new way to significantly reduce charging times by increasing efficiency.

However, mankind will never stop searching for new materials. In addition to graphene, the search for a material that can be used to store high-density electricity in the future will require continuous screening to find out the different characteristics of different materials. In addition, in addition to finding new materials, how to find more ways to improve energy density, reduce costs and seek greater breakthroughs based on existing materials by changing the formula, is also worth exploring the path.

From paper information storage to silicon information storage, mankind has achieved one information storage revolution after another, basically getting rid of the monopoly and limitation of information and breaking the information divide. The key challenge to be solved in the next 50 years or more will be the mass storage of energy, and if this problem is better solved, there will be even more surprising and dramatic changes in human society.

Society's Ethics and Morals Face Dramatic Changes

Historically, the defining forces of a society's philosophy, morality, ethics, thought, culture and religion have been the development and transformation of technology. The development of technology will lead to a change in the material base, which also means the amount of material available for distribution. In other words, in certain material conditions there are philosophies, morals, cultures and ethics that go with them. It can be argued that all human thought cannot transcend technological development and material foundations, and that all theory is limited by material capabilities in a technological context. It is a constant fact that naturally dynamic technologies are forever changing the material world.

The drive to improve technology is inherent in the human being, it is rooted in the human heart and is a necessity for survival and development.

Confucianism arose in the Spring and Autumn period as a result of socio-economic development. The emergence of the dictatorship of Confucianism in the Han dynasty came after a long period of social stability and economic development following the change of dynasties and the urgent need for the ruling class to strengthen centralised rule. The need to maintain rule led to the consolidation of ideas and the taking of what was needed. "Heavenly and human, divine authority" "Heaven is the progenitor of all things, and all things cannot be born without it." man is man, and man is born of heaven. Heaven is also the great-grandfather of man, and this is the reason why man is like heaven above. "The body of man is formed by the number of heaven" the central goal of these ideas was to emphasise the divine right of kings, to unify the kingship with the divine right, to create a religiousisation of the kingship, and to use it to strengthen their own dominance.

Since then, Chinese thought and culture in the millennium has been dominated by Confucianism, which, in addition to serving the ruling class, has also been in tune with the development of agricultural society. With the advent of the industrial revolution, Chinese Confucianism, which had lasted for thousands of years, came under attack, especially against the backdrop of the technological and economic advances of the new era, when foreign ships and cannons broke down the previously closed culture and thought. As a result, Confucianism fell apart. The fundamental cause of this phenomenon was not some remarkable change in thinking, but the emergence of new ideas and cultures, driven by technological change, on a social, economic and military basis.

The first industrial revolution, the use of the steam engine, changed the face of mankind. Steam trains changed people's perception of speed and it was a new experience to sit on a train and see a speeding horse-drawn carriage backing up quickly. At the same time, the steam engine production line, with its devilishly automated rhythm of machine parts, gave a glimpse of the power of technology.

The use of machines has brought about huge changes in human behaviour and thinking, especially as what was previously regarded as truth for a long time is constantly being broken by new things and new rules are being established. For example, as with many historically significant new products, the early cars were only available to the wealthy few and the general public did not have access to them. Over time, researchers found that a single car was not expensive enough to produce to allow more people to benefit from the invention, and a car for dozens of people to ride together—a bus—was born.

The bus is characterised by the fact that several people can ride together and, as a common product, it solves the problem of the general public not being able to use the fruits of a new invention. The advent of the bus has given mankind more insight into how to get more people to use and share a new product together. This new idea has given rise to a new concept: the public good.

In the context of rapid technological development, the material wealth of society has increased tremendously and the relationships between people, nature and society have changed dramatically, with new social relationships beginning to emerge.

One hundred years ago, homosexuality was not tolerated by the majority of people in the region, because reproduction was a basic human requirement, and homosexuality could not fulfil this requirement, and it was not socially acceptable even if the people involved had feelings for each other. With the development of society and the progress of human civilisation, the twenty-first century has seen a change in attitudes towards homosexuality and a gradual acceptance of it by human society.

This progress and change is based on the material development of society, as the importance of human reproduction has declined, based on much higher food production, higher infant survival rates and a better social security system, and the fear that "there are three unfilial children and no children" is not as strong as before. Thus, the human desire to reproduce gives way to love, freedom, understanding and respect.

Material changes are also influencing national policies and laws. In the late 1970s, family planning was widely disseminated and accepted by the social elite in China, and by the early 1980s, family planning was established as a basic state policy. This perception is based on the premise that life expectancy is increasing, infant survival rates are rising and birth rates are increasing, but the level of material production in society is not, i.e. the level of material growth in society is not keeping up with the rate of population growth. As a result, with the exception of the more backward-minded rural areas, the elite of society understood and supported the state policy of family planning.

Today, after more than 30 years of development, family planning policies have been relaxed and are even gradually being encouraged, based on a significant increase in material production capacity and a consequent change in people's perceptions and thinking. For example, thanks to the large-scale use of fertilisers and pesticides, our grain production has increased considerably. In 1980, China produced about 342.5 million tonnes of grain, and by 2016, it had nearly doubled to 616.24 million tonnes; at the same time, the production of agricultural products such as cotton, oilseeds, vegetables, fruit, pigs, cattle, sheep and poultry has increased significantly and at a rate far faster than the growth of the population, with the same amount of land and resources already available to feed more people.

In addition, there has been a spectacular explosion of production capacity in the industrial and construction sectors, driven by new technologies. We have gone from a society of scarcity to one of overcapacity. If there are no more people to consume, the products produced will be wasted and the economy will be in the doldrums. As a result, the call for an appropriate increase in population and a relaxation of the family planning policy is gradually being accepted, and the government is taking stock of the situation and adjusting the policy accordingly.

The change in family planning policy also demonstrates that technological change is the basis for an increase in material production capacity, which in turn leads to a change in policy, ideology and, ultimately, social values.

After 5G Humanity Will Gradually Move Towards Intelligent Communism

After 5G, human society will gradually enter an era of intelligence, which will lead to huge changes in social structure, thought and culture, and will even usher in the era of intelligent communism.

The original human understanding of communism was that public goods could be owned in common and shared by the majority. Engels had devoted his book The Principles of Communism to the principles of communism and the conditions for its realisation: the proletarian revolution will establish a democratic system of state and thus, directly or indirectly, the political rule of the proletariat. It could be established directly in Britain, where the proletarians were now the majority of the people. It can be established indirectly in France and Germany, where the majority of the people are not only proletarians, but also small farmers and petty bourgeoisie, who are in a transitional stage of transformation into proletarians and who are increasingly dependent on the proletariat for the realisation of all their political interests and who will therefore soon agree to the demands of the proletariat. This may require a second struggle, but this one can only end with the victory of the proletariat.

Scientific socialist theory also argues that the speed of development of a communist revolution depends on whether the country has more developed industry, more wealth and a relatively large number of productive forces. This theory asserts that the historical conditions for the achievement of a communist society include the following: the high level of development of the productive forces of society, the extreme development of science and technology, the unprecedented increase in the productivity of labour, the great reduction in labour time and the great abundance of social products; all private labour and small-scale production is replaced by socialised mass production, the means of production are shared by all members of society, and commodities and money disappear; all people are born equal and the differences between workers and peasants, between urban and rural areas and between mental and physical labour are eliminated; as a result of the change in the mode of production and the demise of the old social division of labour, every member of society will be free and fully developed; the whole of society is systematically mass-produced to suit the needs of different people; the principle of a needs-based distribution according to each individual's ability, although the needs-based distribution still depends on the extent to which people are able to contribute; there would be no classes, the "enslavement of men" would be replaced by the "management of things", the state apparatus would die out on its own, but the organs for the punishment of crime would remain; as all systems of economic oppression and slavery die out, so do all the unequal moralities, religions, etc. of class society...

Obviously, scientific socialist theory did not foresee the emergence of an intelligent Internet, much less the emergence of artificial intelligence, nor could it have predicted the impact and influence that the development of an intelligent society would have on human nature and human habits of mind. In fact, with the advent of artificial intelligence, new values and attitudes towards people are being rebuilt. In material production, AI will replace most service jobs as productivity increases and a small number of people are able to perform jobs that previously required many people. For some people, it is possible to live without working, and even for a society, when what is lacking is not productive capacity, but consumption capacity. The question of how human society should face this new change in the future is a major issue before us.

Artificial Intelligence and the Phase of Great Material Abundance Will Come Sooner Than Expected

Whereas in the past, social production activities were carried out by everyone, otherwise society would not have been able to feed a large population, this era is now passing and social production activities are becoming highly centralised and carried out by a few people, a trend that is accelerating even faster than we could have imagined.

The information revolution in the history of mankind has been much shorter in coming than the previous one. The first information revolution, the emergence of language, took place a million years ago; the second information revolution, the invention of writing, took place some 5000 years ago; the third information revolution, the invention of paper and the printing press, took place only 3000 years ago; the fourth information revolution, the invention and widespread use of radio, is only 300 years old; the fifth information revolution, the invention and use of television, is only a century old; the sixth information revolution, the emergence of the Internet, took place only 60 years ago.

With the advent of the Industrial Revolution, mankind's material wealth entered a period of high speed development, where physical distances became shorter and time became faster. In particular, a large number of machines were invented to increase production and labour efficiency, which led to a rapid increase in mankind's possession of food, steel and energy. For example, cars, aeroplanes, ships and other vehicles with high speeds were invented and widely used for transportation; tractors, harvesters, threshers, dryers, seeders and other machinery were used for large-scale agricultural production; the use of fertilisers, pesticides and herbicides in farming, together with scientific selection, breeding and improvement of seeds, has greatly increased the yield of agricultural products.

It is clear that the industrial revolution has greatly increased social wealth, and that the smart society that will follow 5G will allow it to grow even more spectacularly.

The basis of human society is agriculture, and for a long time it was thought that agriculture would be difficult to achieve mass production and industrialisation. The fact is, however, that the forces of industrialisation have now entered agricultural production. In 2017, I went to Beidahuang for research. Before 1947, Beidahuang was basically a wasteland, but after several rounds of development in the 1950s and 1960s, today Beidahuang can produce food for 120 million people, 98% of which is commercial food, which means that the local people can produce a large amount of products with a small amount of food consumption. In the local area, where there are as many as a thousand plots of land in a 10,000 acre block, preferential breeding and large mechanical production have become a reality, with only a handful of farmers operating machinery on the land. In the 1950s, an acre of rice yielded only 300 kilograms; today, 1000 kilograms per acre is the norm. The production process is also much less labour-intensive thanks to machine-based operations.

In the future, the involvement of intelligent technology will further increase the production of human food. For thousands of years, food production has been largely dependent on the weather, with people being unpredictable, less resilient to the weather, and with little knowledge of the land. In the age of intelligence, humans can not only predict the weather, but also intervene effectively to meet the needs of production. In the process of food production, previously unsolvable problems can also be achieved through the machine, for example, in the past encountered rainy weather, a large number of grain will germinate, mould, the future can be completely through the large drying equipment for drying and storage.

In the age of intelligence, every piece of land can be scientifically monitored and managed before cultivation, such as the temperature, water content, trace elements and fertility of the land can be effectively monitored, then targeted analysis can be carried out and fertilisers can be applied according to actual needs to ensure the best conditions for food production, supplemented by preferential breeding and genetic modification, food production and production efficiency will be greatly improved.

Not only food but also vegetables will be grown industrially. In the past, vegetables were first grown in the most primitive way in marginal plots, then gradually shifted to large field production, and now greenhouses are beginning to become the main form of vegetable production. In the future, vegetable cultivation will be replaced by standardised plants, when vegetables will be customised according to market needs, the temperature, light and humidity in the plant can be adjusted to meet the best conditions for vegetable growth. The content of the base fertiliser and various micronutrients can be adjusted in proportion to the needs of the vegetables. In this type of vegetable factory, there are no pests or diseases, so there is no need to use pesticides, safety is increased, production is faster and quality is greatly improved.

It is foreseeable that farming will also go factory. The shad, for example, is a migratory fish that must live in live water and is extremely difficult to breed in captivity. In the case of intelligent factory farming, this rare species can be farmed on a large scale by simulating the living environment of the shad for nursery purposes and then simulating the living environment of the shad to establish migratory flows.

Today, the vast majority of countries have their basic food needs met and are moving towards higher quality and greater safety. With the advent of the smart age, the food problem will be better solved and the number of people involved in food production will be drastically reduced. In the next 50 years, the peasant profession will gradually decline, the traditional peasant will no longer exist and the work of peasants will be gradually replaced by agricultural labour. This process will be accompanied by a decline in the countryside and its disappearance. The transformation of society will also lead to a large number of peasants moving into the cities, and the proportion of agricultural workers in the total population will become lower and lower, so that food production and management will require only a very small number of people, from seeds, planting, harvesting to marketing, and big capital will be involved, forming an efficient system.

In the post-5G era, intelligence will also make industrial production more efficient and systematic, and industrial workers will be significantly reduced, most typically in the field of electronics manufacturing.

There were once hundreds of mobile phone companies around the world, but today they have shrunk to a few dozen, mainly in China, and each company is very concentrated, both in terms of brands and OEMs. At present, there are fewer and fewer workers on the production line due to the increasing level of intelligence. In 2015, Huawei's Songshan Lake production base had 128 workers on one production line, which was reduced to 28 in 2017 and to 19 in 2018.

From the point of view of industry concentration, the majority of mobile phones used around the world come from a few factories, not only the whole machine is gradually concentrated, accessories are also concentrated to a small number of enterprises, while the bottom of the chip, can do design and packaging, testing few enterprises. Among them, the most basic lithography machine, only a few companies can produce. It was previously unthinkable that a few or a dozen companies could produce a product that could be used by billions of people worldwide. As for other companies, they have completely lost the ability to compete with each other because of various problems such as efficiency.

In industrial production after 5G, the process of intelligence will be greatly accelerated and the word "intelligence" will play an even more important role. In the case of air conditioning, for example, the traditional function is simple: it can provide cooling or heating to meet daily needs. With the development of technology, functions such as dehumidification and energy saving were added later, but these capabilities are independent and air conditioners can be used as long as they have these functions. The future era of intelligence will turn products into a service system, the same brand or the same platform of products, in addition to air conditioners, there will be air purifiers, dehumidifiers, humidifiers, heaters, these products are linked with instruments with environmental monitoring functions, in the case of full analysis of the environment, intelligent decision-making, to meet a variety of needs. At that time, if companies produce air conditioning products too single function, the user experience will be greatly reduced, and the ability to provide a wealth of services to the enterprise, only a few. For a country, there will not be many platforms producing the same product and many small businesses will lose out.

The social services sector will also change dramatically in the age of intelligence and the demand for service workers will inevitably decrease. Retail was once one of the largest and most segmented industries, which had to penetrate every corner of society, with countless small shops and vendors involved in retail activities. Since then, the supermarkets of the industrial age have squeezed most small shops out of existence. In the age of intelligence, e-commerce, especially mobile e-commerce, has become popular, with several huge platforms replacing many small shops, unmanned sales, electronic payments, drones and unmanned vans entering all aspects of sales, more diversified transport of goods and the possibility of delivery at agreed times. Once such a system is established, the once large and fragmented retail industry will be centralised, from procurement, distribution, sub-assembly to the final delivery to the hands of the user, intelligence will allow a significant increase in efficiency, the operator's costs will be greatly reduced, the past can be seen everywhere a large number of practitioners will gradually exit the industry.

Intelligence will also allow a large number of robots to be used in hotel service and restaurant service. For standard tasks such as making coffee and clearing tables, robots will soon be able to do them with ease, and employees in these areas will be withdrawn.

In short, the great material abundance that we have been waiting for in the past may arrive faster than we expect, and the fundamental force that will lead to this outcome is artificial intelligence.

A Low-Desire Society Makes It Possible to Distribute According to Need

For a long time, people have been sceptical about the distribution of goods according to need, because even with great material abundance, there is no limit to human desire. Does the distribution of needs mean that everyone has to live in a villa and have a Porsche? In fact, it is impossible to satisfy this distribution. Scientific socialist theory also holds that the distribution of the products of labour is based solely on the needs of the human being, and not on the arbitrary satisfaction of any desire of all. More importantly, with the great abundance of material goods, it will not be too difficult for ordinary people to acquire them, and mankind will have a low-desire society.

Desire is an animal instinct, a longing caused by need, a yearning for satisfaction in the context of demand. The greatest incentive for desire is shortage, one of the most primitive instincts of human nature. In the human heart, it is the desire for something that is desired but not available that stimulates desire. If you want something there is something, it is difficult to desire something when you can get it when you want it, or when you can get it without much effort.

In the 1950s and 1960s, Chinese people had a high desire for material things because they had been starved since childhood and had an instinctive fear of hunger, while after the reform and opening up, there was a certain psychological gap in the face of the outside world, and a strong urge for material things was formed inside. For this generation, it was difficult to understand low desires at the time, which explains that desires are potentials caused by shortages and disparities.

In times of material shortage, there is a high demand for material possessions. For example, if eggs are available by ticket, people will not only buy all the eggs they can buy by ticket, and they will also stock up on as many eggs as possible. Since then, as the economy has been revitalised and material goods have become more abundant, the demand for eggs has been fully met and eggs are readily available, so people don't buy more eggs, they just buy as many as they can eat, because the anxiety has disappeared.

The level of desire is inversely proportional to the degree of material satisfaction; the more scarce, the more likely it is that desire will arise, and the easier it is to be satisfied, the greater the reduction in desire, so that a low-desire society can be expected in the future.

With the great abundance of material goods, people will have less difficulty in obtaining the basic necessities of life, and human society will begin to face the test of low desire. In more economically developed countries and regions, such as Northern Europe and Japan, low desire is becoming a new problem.

Generally speaking, a low-desire society is characterised by three basic features:

(1) Fewer children. Not having children is a key sign of a low-desire society. In the past, the concept of having more children and raising children for old age was deeply rooted in people's minds, and the notion that having children was a way to prosper the family and provide more security in old age drove human reproduction. Nowadays, the number of people who do not want to have children has started to increase, and this is due to two reasons: on the one hand, the cost of having and raising a child is too high, parents pay more and bear more pressure, and for many families, having an additional child can cause a decrease in happiness and affect the quality of life. In this context, having only one child, or even not wanting to have children, has become the choice of many people of childbearing age. On the other hand, the social security system is becoming more sophisticated, so that when they grow old, they do not need the care of their children and can spend their old age comfortably in nursing homes and other social care institutions.

The problem of childbearing is becoming a major problem in developed countries, and is becoming increasingly evident in economies such as Northern Europe and Japan, which entered the advanced stage in the 1970s and 1980s. Nowadays, China is also becoming a country with fewer children.

(2) Social entrenchment. Social entrenchment is also a key sign of a low-desire society. With long-term political and economic stability, society will gradually solidify. With the high cost of entry into the mainstream of society for the underclass, there are fewer and fewer opportunities for upward mobility for ordinary people, and the incentive

to work hard is diminished. The biggest problem of class solidification is in education. Not only do people at the top have more resources and a stronger economic base, but more importantly, they invest more in their children's education than other classes, which results in better educational opportunities and resources for those at the top, while those at the bottom do not have better educational resources and development opportunities, thus discouraging them from moving up the social ladder. In other words, the rich get richer and the poor get poorer, once the whole society loses its dynamism, the spirit of innovation will be lost and it will look like stagnant water.

Social consolidation is already occurring in many countries around the world, although China has not yet reached this stage, however, this phenomenon is a cause for alarm and should be addressed well in advance.

(3) Gnawing is prevalent. Once the basics are met, some people lose their motivation and drive, and without the pressure of living beyond their means, they begin to retreat into their inner world, not wanting to face too much competition, not working, not marrying, not having children, and relying only on their parents for survival. Nowadays, this phenomenon is already present in societies with better social welfare and less human relations. With the increasing pace of social development, such people have de facto lost the basic ability to compete in society and have to live on social welfare, and the better social welfare accepts and tolerates the presence of such people.

A high welfare society provides for the survival of those who have lost their motivation, while with rapid technological advances it is possible that a large number of ordinary jobs can be replaced by robots, and that for some people society no longer needs them to work, or that their work creates little value and society can afford to support them.

Today, in developed countries and regions such as Scandinavia, Switzerland and Japan, the characteristics of a low-desire society are becoming increasingly apparent, with a polarisation between the social elite and the general population, in which the general population is losing its motivation, the avenues for upward social mobility are narrowing and there is little pressure for social survival. In terms of basic material needs, the distribution of needs is approaching to a certain extent. Instead of caring about technology, inventions, manufacturing and production, and seeking new opportunities for development, many people are turning their energies to sports, games and entertainment, and fun is becoming an important part of their lives.

Labour Becomes the First Necessity

After 5G, artificial intelligence will be used in a large number of social production, productivity will be greatly increased, most common household goods may be produced by a few large companies for global use, and many services and a large number of jobs in our daily lives can be replaced by robots. The intelligent society will gradually divide into two kinds of people: the working man and the recreational man.

The working man is a minority in an intelligent society, a man who does not do the work of an ordinary man, but works on algorithms, inventions, technological breakthroughs, the construction of models, the search for new materials, the exploration of new methods. An intelligent society places great demands on the people who work there, such as a high level of intelligence, a strong scientific background and a high level of stress tolerance. This type of person either has a profound study in a certain field and can delve very deeply into it, being an expert in a certain field; or has a broad vision, a deep understanding of the future, patterns, the universe and other aspects of the relationship, and has the ability to plan far ahead. Of course, workers are also under pressure because they are constantly learning to improve intellectually and to improve themselves at work.

Entertainers will gradually become the majority of society. At present, a large number of ordinary jobs will be replaced by robots. In the past, many people were engaged in agriculture in order to produce food for all, but as agriculture becomes smarter, food production is mechanised, vegetables, livestock and aquaculture are factory farmed, and the vast majority of the agricultural population moves to the cities, only a small part of society will be engaged in agriculture, and there will be a large surplus of labour.

Not only will agricultural production be replaced by machines, but logistics and transport systems will also not require much human maintenance, with machines playing a major role. In the future, only a few parts and maintenance will need to be carried out by people, and even some maintenance work will be carried out by robots.

It is foreseen that machines will be responsible for most aspects of the production of industrial products. Today, unmanned factories are already present in many fields, and the more technical and sophisticated the field of production, the more it will move towards large machine production, where manual production means high costs and only very few fields, such as art with a personal touch, still need people to do it. The core products in most fields are produced by only a few or a dozen companies worldwide, products tend to be standardised, production costs are lower, efficiency is higher, there will be a significant reduction in the number of general industry workers, and manufacturing will also become a minority affair.

Intelligence will also play a major role in the service sector, with urban transport consisting of driverless cars, high-speed railways and capsule trains forming a complete intelligent transport system, the system will be controlled and managed by an intelligent transport system, which will no longer require a large number of people to manage and control the system, with drivers and attendants, as well as most other general operators, retired.

Logistics, intelligent delivery robots, intelligent drones to replace the daily logistics system, takeaway and online shopping goods will be delivered to your doorstep through the intelligent delivery system, in such a system, the importance of the current delivery man declined, the courier boy to switch to become inevitable.

Those who leave the general workforce will move into the entertainment industry, providing entertainment for the community. At the same time, many people in society no longer need to work, so entertainment has become a part of their lives, in which they can transform themselves into any role in society and find great physical and mental satisfaction.

For the working man, his reward is not wealth or material reward, when the wealth of society has increased so much that the acquisition of material wealth is no longer so important, and there is no great difference between a rich man and an ordinary man in terms of the material he gets and the material he consumes. In the age of intelligence, the needs of survival, security and emotion have been addressed. The reward for the workforce is control and social recognition. The highest level of human needs is the need for respect and self-fulfilment.

In such a world, it was a luxury and a source of pride to be a working person, to be involved in labour, because at that time the average person did not have the opportunity to be involved in work and more people were only entertainers, and it was difficult for them to be respected and to achieve self-fulfilment. In the age of intelligence, therefore, labour becomes the first necessity and becoming a worker will be an important sign of moving to the top of the social pyramid.

Intelligent communism is the idea that when human society enters an intelligent society, human thinking will be very different from that of the shortage era. With the great abundance of material goods, the needs of the average person are easily met and basic resources can be distributed according to people's needs. In this state, mankind enters a low-desire society, where the pursuit of material goods is greatly reduced, where most people do not need to work to obtain basic subsistence, where labour becomes a realm that is difficult for the vast majority of people to attain, and where only a few people are involved in work, where work becomes a symbol of social respect and self-fulfilment.

In a situation of great material abundance, where a large number of jobs are replaced by machines, human nature will change dramatically, philosophies, morals, ethics, culture and customs will be very different from today, people's perceptions of money, their views on material things and their relationships with each other will become very different, the fears caused by material shortages will be eliminated and the sense of well-being will increase. At the same time, people become more curious about the universe, the future and other unknown worlds. And the way humans solve their problems will move more from war to negotiation, because on the one hand they do not need to fight for resources to solve their survival problems, on the other hand, wars in the age of intelligence are likely to be much more violent and destructive, and the cost to mankind will be unbearable if they start lightly.

A New Human Species of "Intelligent Human" to Emerge

For a long time, mankind has imagined a future world of intelligence in which robots will triumph over humans and where humans and robots are the antagonists. From the human point of view, it is hoped that robots will always be under human control.

With the development of intelligence, a large number of robots may have the ability to think and even to reproduce, and robots will certainly play an important role on Earth, with powerful and intelligent robots becoming part of the planet.

I also believe that between the carbon-based human and the silicon-based robot, there will be a new species of human, an intelligent human that combines the carbon-based and the silicon-based.

While we marvel at the strength and increased intelligence of robots, we must also see the value of humans as carbon-based animals, who get their energy from bioenergy, which can be regenerated continuously on earth, from animals and plants, and from a variety of sources. The human brain is a bio-computational storage system, which is a very low-energy system, with the adult brain consuming only 250 kcal–300 kcal of energy a day. In other words, a brain weighing between 1300 and 1400 grams (the average weight of an adult brain) has a power of about 15 watts, consumes very little energy, performs complex calculations and stores, and does not heat up, storing texts, pictures, sounds, images and recording logical relationships. The human brain has 100 billion nerve cells, and it is not known today whether the functions of these cells can be fully exploited for computing and storage, but there is certainly a large potential for the brain to be exploited.

The brain has a very scientific mechanism for processing information, and all the different information in the brain is stored and recalled at the same priority, in fact, the brain scientifically manages the information it receives over time and in accordance with its importance, storing it in different priorities. Information that is older can also be compressed, blocked or even deleted so that the brain is not so overloaded, leaving the computing and storage capacity to deal with high priority issues at all times. The information that lies deep in the brain can also be called up and awakened through a hypnotic mode. Today's research on the human brain has not yet fully solved the mystery and understood all its mechanisms.

Beyond the brain, there is a great deal that remains to be understood about the genetic inheritance of information in humans. When a child reaches the age of 2 to 3 years old, he suddenly learns to speak, and not only speaks, but also understands a large number of logical relationships that adults have not taught him, or have taught him, and which would be difficult for a child of 2 to 3 years old to understand, but the fact is that he understands them. The human understanding of the world, especially the various logical relationships, is similar to a computer algorithm, the most basic logical relationship, typed into the code of life, which is passed on to the new life through RNA, and when the child grows to a certain stage, in the face of suitable external environmental stimulation, these codes of life will be opened. The child's language skills are fundamental to these codes of life and, when combined with impressions of external stimuli, can easily replicate the most basic knowledge and understanding of the world of previous generations.

There are many channels of dark matter in the transmission and storage of human information that we have not discovered, and the human species is by no means as simple or as easily replicated as robots.

Will the capabilities of the human brain be exploited more in the future? There is a greater opportunity in this area. And it is a certainty that humans will transform themselves as their technological capabilities increase. For example, the connection between the neurological system of the human brain and the chip is made by implanting a chip. Initially, the chip would only detect changes in brain waves, making auxiliary judgements and driving the rest of the human nervous system. The more complex chips will gradually become integrated with the human brain, and the information stored in the human brain can be copied to the implanted chips, and the information from the implanted chips can be transferred to the human brain through copying, so that human beings can no longer need to learn the solidified knowledge, and do not need to strengthen the memory by reciting it over and over again to solidify the storage, but can directly store the information in the chips and recall it.

In this case, the human learning rate will be greatly increased, the learning effect will be enhanced, and some solidified knowledge will not need to be learned.

At the same time, humans could transfer copies of information from their brains to chips, so that we would not have to worry about the irreversible death of brain cells, and knowledge and memories could be stored on the chips. This is the highest level of future artificial intelligence, and when this technology is realised, mankind will enter a whole new era.

In addition to the brain, humans can also modify their external skeleton. This technology is already well developed today, with the addition of more artificial intelligence and new materials, exoskeletons will be stronger and lighter, and the joints will be more flexible. These exoskeletons, together with artificial intelligence, can greatly increase a person's ability to carry weight, run, jump and other activities, and can take the place of organs such as arms, legs and feet when they fail, or even support a paralysed person. These exoskeletons become an aid to human capabilities and will raise human mobility, movement and weight-bearing capacity to a new level that many human muscles and bones are unable to withstand.

New materials and artificial intelligence can recreate most of the human organism and enable long working hours. Today there are already devices such as pacemakers that help the heart to work, and with the discovery and perfection of artificial intelligence, microbattery technology, bioelectricity and new materials, most of our human organs can be replaced and reconstructed, the heart, blood vessels, skin and many other organs can be replaced. Through organ reconstruction, humans can live longer and be more capable.

One day, there will be a person whose organs have been rebuilt to be far more athletic and weight-bearing than the average person, and who will have a chip implanted in his body that will connect him to the brain's nervous system, greatly enhancing his ability to learn and store knowledge, reacting extremely quickly, and processing problems much better. Naturally he is not a robot, he is still a carbon-based human being, he will fall in love, get married, have children and reproduce normally like a human being, but the intelligent transformation makes him completely different from a human being, his understanding of things and his feelings will also change. From this point of view, are such people still human? How will we relate to them? Will we choose to be like that?

It may be that for a long time humans will not be able to accept that they have been transformed, but in the course of time this problem will surely be broken down little by little. When human life is at stake, someone will try to do so. And as many people make this choice, more and more people will become intelligent, with capabilities far beyond those of ordinary humans. How we accept them, understand the value of such people philosophically, accept their existence in all aspects of social life, and safeguard their rights and fairness in law are complex issues. But even if it is complex, humanity may have to face up to the existence of such a group, which combines carbon-based and silicon-based people.

Facing the Universe Will Be the Theme of the New Era

The vastness of the universe will always be the aspiration of mankind. It is vast, life-giving and full of mystery. For a long time, mankind has looked up to and imagined the universe, and has done very little to explore it.

In the age of 5G and beyond, the most mysterious frontier for mankind to explore is the distant universe. Humanity has solved the problem of production and resources through artificial intelligence, the population is no longer expanding at a high rate, basic survival is not a problem, diseases are no longer completely invincible, and wars can be dissipated through communication. At that time, the most challenging thing facing mankind will be the exploration of the universe. I believe that, like the age of great maritime discoveries in history, the age of great cosmic discoveries will come.

The advent of the Age of Sail was made possible by improvements in shipbuilding technology and the development of astronomy, which allowed mankind to overcome the turbulent seas and go out to sea in search of opportunities and treasures.

In the age of cosmic discovery, first and foremost, spacecraft had to have the capacity to travel long distances and withstand long journeys. Such a spacecraft is definitely not a small space shuttle or lunar module, but a complex system that supports multi-person life, can be resupplied for long periods of time and can travel long distances. This system, which needs to be strong and powerful enough to withstand all kinds of impacts and collisions, is also self-healing.

Energy supply will be a big problem to be solved for cosmic voyages. In the solar system, solar energy can be collected through solar panels, but after flying out of the solar system, the challenge of obtaining a continuous supply of energy in the face of the distant universe is an extremely complex one. Perhaps the absence of sunlight for much of the flight necessitated the development of a more powerful energy system that would allow the vehicle to travel longer distances. This requires a revolutionary breakthrough in the generation and storage of energy, and it is crucial to find a more powerful mode of energy generation than nuclear energy, and to store electricity that can be sustained over long periods of time.

The material is also a major difficulty. Spacecraft need to fly in ultra-high and ultra-low temperatures and are exposed to a wide range of impacts and stresses, which require the lowest possible energy consumption. In addition, materials for spacecraft need to be high density, high strength, lightweight, radiation, penetration and corrosion resistant in order to cope with the special cosmic environment.

How to achieve high-quality communication in the age of cosmic exploitation is also a challenge to be solved. Maintaining communication with the Earth, being able to fly away from it and one day back, and maintaining the exchange and transmission of information between the craft and the Earth at all times, will test human ingenuity.

To combat the barriers of time and ageing that carbon-based humans cannot cross, humans are frozen or put into hibernation so that they do not age during long flights, and memories are backed up on hard drives so that they are not lost as brain cells lose their vitality and are restored to their minds when they reach their destination.

In the future, mankind will be able to set up multiple interstellar staging posts and move on to other planets, even beyond our solar system and into the galaxy. Humanity's exploration of the universe requires not only the resolution of a large number of technical problems, but also philosophical and ethical challenges, such as what is the place of humanity in the universe in the future? Does the human world belong to the earth or to the universe? How do humans and life on other planets in the universe get along?

There are many boldly envisaged questions that we cannot answer today, nor can we, but in the not too distant future we must answer them and will certainly find the key to their solution.

5

Conclusion

Any technological revolution drives human progress. Today, technological progress has entered an era of acceleration, and each generation of mobile communication brings not only changes in the communication technology itself, but also changes in the industries associated with it, affecting not only the technology, but also the industries, products and services, which in turn affect the economy, society, culture and ultimately ethics, morality and philosophy.

For a long time, society as a whole has only understood 5G in terms of speed. Of course, 5G is first and foremost about speed. As tariffs continue to fall and a large number of users join, today's 4G networks are no longer able to meet user demand, with download speeds dropping from 50Mbps to 5Mbps in many parts of China, creating a need for a 5G network with greater capacity, higher spectrum utilisation and a better experience.

The greater value of 5G is to see communication networks in a new light. This network is no longer about human interaction in the traditional sense, nor is it just a network of Internet access, but through this network, machines start to interact and talk with each other, the terminals running in this network are no longer mobile phones operated by people, and there is no longer a person behind each terminal, the concept of user may gradually withdraw, because each user may have several or even more terminals.

L. Xiang, *The 5G Era*, https://doi.org/10.1007/978-981-99-4534-4_5

In the 5G era, cars, parking spaces, utility poles, manhole covers, cameras, door locks, washing machines, environmental protection, air purifiers, range hoods, heating controls, refrigerators and other devices will all be connected to the Internet, and mobile communication users will skyrocket from an average of one mobile phone per person today to an average of more than 10 terminals per person. This explosive power will not only create a huge market in an industrial sense that has never been imagined before, but will also lead to huge industrial opportunities. Initial estimates suggest that by 2025, there will be 10 billion mobile terminals in China, not primarily mobile phones, but a multitude of IoT devices. 5G's low power consumption, low latency and the ability to connect everything makes this huge market possible.

5G is also bound to bring dramatic changes to the industry. The previous kind of simple network will become a multi-slice intelligent network, telecom operators will also be from the past network construction and operation, management billing, etc., facing a more massive management system construction, billing system construction, network operation will become a more complex problem.

In the business sector, 5G will certainly open up new opportunities and some disruptive services will emerge. In the 4G era, mobile e-commerce and mobile payments have suddenly found their breaking point. These applications, such as takeaways and taxi rides, have benefited from the wider coverage, higher speeds and more accurate positioning of the 4G network, which has led to new business models and new business opportunities.

5G offers not only higher speeds, but also low power consumption, low latency and the ability to connect everything. These capabilities will enhance the Internet of Things, big data and intelligent learning, allowing these individual capabilities to form a new convergence effect.

5G will make the video business more vital, the development of direct broadcast, HD video distribution, live streaming business will be further accelerated, advertising, information dissemination, news business will take different forms because of 5G. 5G will also make m-commerce completely different from what it used to be.

It is foreseeable that 5G will soon push the smart home, which has been promoted for 20 years but not really developed, into a new realm.

More importantly, 5G will penetrate into the public management system of society, making it easier to manage smart traffic, smart cities, telemedicine, smart health management, pollution management, disaster monitoring, etc.

By then, 5G will penetrate from life services to production management: flexible smart factories need 5G, smart logistics system needs 5G, smart agriculture needs 5G…

In the next 5 to 10 years, with the large-scale deployment of 5G networks, human informatisation will enter a new era, in which communication will no longer be simple communication, but a new system integrating communication with intelligent sensing, big data and intelligent learning. Under the new system, new business models, commercial models and service models will create huge new opportunities, which will be an important force in driving the economy in the future. In the landline era, telephony was defined by population and there were no more than 300 million landline users in China. But in the age of mobile phones, because each person can have more than one mobile phone, today there are more than 1.4 billion mobile phones in China. In the future, there will be several or even more smart devices in each household and each person, and social service devices will also average one per person. In this context, the probability of an eruption of smart Internet terminal devices is higher.

It is difficult to imagine the market and opportunities of the mobile Internet when you look at it with the traditional Internet mindset, which can create more market opportunities and business models different from the former. The smart Internet built on 5G will not be a simple copy of the mobile Internet, a copy of business and business models, but will be based on communication, sensing and artificial intelligence, transforming social management and social services, and truly entering a smart living society.

In the 4G era, China has built the world's most comprehensive network—more than half of the world's base stations are in China—and has successfully driven the development of China's mobile Internet business, with mobile electronic payments, mobile e-commerce and takeaway services at the forefront of the world, forming many of the world's leading new mobile Internet businesses. In the 5G era, China will also become the world's leading country in 5G construction, and the construction of 5G will not only improve communication capabilities, but also fundamentally change the ability of social management and social services. With the arrival of 5G, China and the world will change dramatically and we look forward to the day when this will happen.

References

Yuval Hurali. A brief history of mankind: from animals to God [M]. Lin, Junhong. Beijing: CITIC Press, 2014.

Hu Zhuanglin. A Course in Linguistics [M]. Beijing: Peking University Press, 2007.

Wu Jun. The Age of Intelligence: Big Data and the Intelligent Revolution Redefine the Future [M]. Beijing: CITIC Press, 2016.

Jingdong R&D System. Jingdong technology demystified [M]. Beijing: Electronic Industry Press, 2014.

Li Lupeng, Xiong Shangkun, Wang Qingyang. 5G technology outlook [C]. Chinese Communication Society. 2013 National Wireless and Mobile Communication Academic Conference Proceedings (on). Qingdao: China Academic Journal Electronic Journal, 2013:14–16.

Wang Wei. Study on the optimization method of CDN content distribution network [D]. Wuhan: Huazhong University of Science and Technology, 2009.

Jin Gulun. Information superhighway and cultural development [J]. China Social Science, 1997 (1):5–6.

You Xiaohu, Pan Zhiwen, Gao Xiqi, et al. 5G mobile communication trends and some key technologies [J]. China Science: Information Science, 2014, 44(5): 551–563.

L. Xiang, *The 5G Era*, https://doi.org/10.1007/978-981-99-4534-4

Zhao GF, Chen JJ, Han YuanBing, Xu Chuan. A review of key technologies for 5G mobile communication networks [J]. Journal of Chongqing University of Posts and Telecommunications (Natural Science Edition), 2015, 27(4):441–452.

Qiao N, Lu Y. TD-SCDMA is the right biography [J]. Communication World, 2006(3).

He Jing, Chang Jiang. Self-organizing network (SON) technology and the evolution of standardization [J]. Post and Telecommunications Design Technology, 2012(12): 4–7.

Hu Po, Li Wenyu, Song Aihui. Advances in self-organizing network technologies and standards [J]. Telecommunications Network Technology, 2012(12): 53–57.

Qian Zhihong, Wang Xue. A review of D2D technologies for 5G communication networks [J]. Journal of Communication, 2016, 37(7):1–14.

Zhou Daiwei, Wang Zhengya, Zhou Yu, et al. 5G terminal service development trend and technical challenges [J]. Telecommunications Network Technology, 2015(3):64–68.

Xiang Hongyu, Xiao Yangwen, Zhang Xian, Park Zhuying, Peng Mugen. 5G edge computing and network slicing technology[J]. Telecommunications Science, 2017, 33(6):54–63.

Xu Yang, Gao Gongying, Wang Lei. An analysis of 5G mobile network slicing technology [J]. Post & Telecom Design Technology, 2016(7):19–22.

Ren YG, Zhang L. Outlook of the fifth-generation mobile communication system [J]. Information Communication, 2014 (8):255–256.

Wang J, Li M D. SDN/NFV - opportunities and challenges [J]. Telecommunications Network Technology, 2014(6):30–33.

Zhao Huiling, Shi Fan. The development and challenges of SDN/NFV[J]. Telecommunications Science, 2014(8):13–14.

National Radio Monitoring Centre. History of radio development [EB/OL]. [2010-09]. http://www.srrc.org.cn/news155.aspx.

Ministry of Industry and Information Technology of the People's Republic of China. Economic operation of the communications industry in the first half of 2018 [EB/OL]. [2018-07-19]. http://www.miit.gov.cn/n1146312/n1146904/n1648372/c6265909/content.html.

eFeverEngineer.com. Three scenarios and six basic features and key technologies of 5G [EB/OL]. [2018-05-31]. http://www.elecfans.com/tongxin/20180215636413.html.

Pengpeng Li, Fang Fang, Jeng Peichuan, Zheng Na. Overview of global 5G spectrum research and enlightenment [EB/OL]. [2017-09-12]. http://www.srrc.org.cn/article18863.aspx.

Diexin Internet. The development of modern mobile communication technology [EB/OL]. [2017-08-11]. http://www.sohu.com/a/163575997_99968711.

Walter J. Ong. Ramus, Method, and the Decay of Dialogue [M].Cambridge, Mass.: Harvard Univ. Press, 1958.

Peng Tao, Lu Qianxi, Wang Haiming, et al. Interference Avoidance Mechanisms in the Hybrid Cellular and Device-to-Device Systems [C]. Personal Indoor and Mobile Radio Communications. Tokyo: IEEE, 2009:617–621.

Shao YL, Tzu HL, Kao CY, et al. Cooperative Access Class Barring for Machine-to-Machine Communications [J]. IEEE Wireless Communication, 2012, 11(1):27–32.

Ferrusr, Sallento, Agustir. Interworking in heterogeneous wireless networks: comprehensive framework and future trends [J]. IEEE Wireless Communication, 2010, 17(2):22–31.

IMT-2020(5G) Promotion Group. 5G Vision and Requirements, white paper [EB/OL]. [2014-05-28]. http://www.IMT-2020.cn.

CISCOI. Cisco visual networking index: forecast and methodology 2014–2019, white paper [EB/OL].http://www.cisco.com/c/en/us/solutions/collateral/service-provider/ip-ngn-ip-next-generation-network/white_paper_c11-481360.html.

4G AMERICAS. 4G Americas' Recommendations on 5G Requirements and Solutions, white paper [EB/OL]. [2014-10-23]. http://www.4gamericas.org.

Kolibri 360 GmbH
Ferdinand-Jühlke-Straße 7
99095 Erfurt, Deutschland
produktsicherheit@kolibri360.de